ECONOMIC COMMISSION FOR EUROPE
Committee on Environmental Policy

ENVIRONMENTAL PERFORMANCE REVIEWS

AZERBAIJAN

UNITED NATIONS
New York and Geneva, 2004

Environmental Performance Reviews Series No.19

NOTE

Symbols of United Nations documents are composed of capital letters combined with figures. Mention of such a symbol indicates a reference to a United Nations document.

The designations employed and the presentation of the material in this publication do not imply the expression of any opinion whatsoever on the part of the Secretariat of the United Nations concerning the legal status of any country, territory, city or area, or of its authorities, or concerning the delimitation of its frontiers or boundaries.

UNITED NATIONS PUBLICATION
Sales No. E.04.II.E.2
ISBN 92-1-116888-0
ISSN 1020-4563

Foreword

Environmental Performance Reviews for countries-in-transition were initiated by Environment Ministers at the second "Environment for Europe" Conference in Lucerne, Switzerland, in 1993. As a result, the UNECE Committee on Environmental Policy decided to make the Environmental Performance Reviews a part of its regular programme.

Ten years later, at the Fifth Ministerial Conference "Environment for Europe (Kiev, 21-23 May 2003), the Ministers confirmed that the UNECE programme of environmental performance reviews (EPR) had made it possible to assess the effectiveness of the efforts of countries with economies in transition to manage the environment, and to offer the Governments concerned tailor-made recommendations on improving environmental management to reduce their pollution load, to better integrate environmental policies into sectoral policies and to strengthen cooperation with the international community. They also reaffirmed their support for the EPR programme as an important instrument for countries with economies in transition, and decided that the programme should continue.

Through the Peer Review process, Environmental Performance Reviews also promote dialogue among UNECE member countries and harmonization of environmental conditions and policies throughout the region. As a voluntary exercise, the Environmental Performance Review is undertaken only at the request of the country itself.

The studies are carried out by international teams of experts from the region, working closely with national experts from the reviewed country. The teams also benefit from close cooperation with other organizations in the United Nations system, including the United Nations Development Programme, the United Nations Environment Programme, the World Bank and the World Health Organization, as well as with the Organization for Economic Cooperation and Development.

This Environmental Performance Review is the nineteenth in the series published by the United Nations Economic Commission for Europe. I hope that this Review will be useful to all countries in the region, to intergovernmental and non-governmental organizations alike and, especially, to Azerbaijan, its Government and its people.

Brigita Schmögnerova
Executive Secretary

Preface

The Environmental Performance Review (EPR) of Azerbaijan began in March 2003, with the first preparatory mission, during which the final structure of the report was established. Thereafter, the review team of international experts was constituted. It included experts from Canada, Croatia, Denmark, and Georgia, together with experts from the secretariats of the United Nations Economic Commission for Europe (UNECE), the Regional Office for Europe of the United Nations Environment Programme (UNEP/ROE) and the European Centre for Environment and Health of the World Health Organization (WHO/ECEH).

The review mission took place from 8 to 21 June 2003. A draft of the conclusions and recommendations as well the draft EPR report were submitted to Azerbaijan for comment in September 2003. In October 2003, the draft was submitted for consideration to the Ad Hoc Expert Group on Environmental Performance. During this meeting, the Expert Group discussed the report in detail with expert representatives of the Azerbaijan Government, focusing in particular on the conclusions and recommendations made by the international experts.

The EPR report, with suggested amendments from the Expert Group, was then submitted for peer review to the UNECE Committee on Environmental Policy on 20 October 2003. A high-level delegation from the Government of Azerbaijan participated in the peer review. The Committee adopted the recommendations as set out in this report.

The report covers twelve issues of importance to Azerbaijan, divided into three sections, including the framework for environmental policy, management of pollution and natural resources and economic and sectoral integration. Among the issues receiving special attention during the reviews were the policy, legal and institutional framework, public participation in decision-making and access to information; the use and supply of water resources, including drinking water; land use, agriculture and biodiversity; management of waste and contaminated sites; selected Caspian Sea issues and environmental concerns in the oil and gas sectors. The report notes that, during its twelve years of independence, Azerbaijan has steadily improved its system of environmental protection and that much progress has been made. In 2001 it established a Ministry of Ecology and Natural Resources, and this Ministry has been leading a strong effort to make environmental management a priority within the country.

The UNECE Committee on Environmental Policy and the UNECE review team would like to thank both the Government of Azerbaijan and the many excellent national experts who worked with the international experts and contributed with their knowledge and assistance. UNECE wishes the Government of Azerbaijan further success in carrying out the tasks before it to meet its environmental objectives and policy, including the implementation of the recommendations to support and promote environmental protection, to improve overall national living standards, and to strengthen international cooperation.

UNECE would also like to express its deep appreciation to the Governments of Denmark, Germany, Hungary, the Netherlands, Sweden, Switzerland and the United Kingdom and to the European Commission for their support to the Environmental Performance Review Programme, to the European Centre for Environment and Health of the World Health Organization for its participation in the Review, and to the United Nations Development Programme, the World Bank and WHO for their contributions to the work in Azerbaijan and the preparation of this report.

LIST OF TEAM MEMBERS

Ms. Mary Pat SILVEIRA	ECE secretariat	Team Leader
Mr. Ivan NARKEVITCH	ECE secretariat	Project Coordinator
Mr. Jyrki HIRVONEN	ECE secretariat	Introduction
Mr. Zaal LOMTADZE	Georgia	Chapter 1
Mr. Mikhail KOKINE	ECE secretariat	Chapter 2
Mr. Mikhail KOKINE	ECE secretariat	Chapter 3
Mr. Antoine NUNES	ECE secretariat	Chapter 4
Ms. Mijke HERTOGHS	ECE secretariat	Chapter 5
Mr. Ivan NARKEVITCH	ECE secretariat	Chapter 6
Mr. Jörgen Bygvraa HANSEN	Denmark	Chapter 7
Mr. John CARSTENSEN	UNEP	Chapter 8
Ms. Stella SATALIC	Croatia	Chapter 9
Mr. Bo LIBERT	ECE secretariat	Chapter 10
Mr. Niels Juhl THOMSEN	Denmark	Chapter 11
Ms. Francesca RACIOPPI Mr. Colin L. SOSKOLNE Ms. Hiroko TAKASAWA	WHO/ECEH	Chapter 12

UNECE Information Unit
Palais des Nations
CH-1211 Geneva 10
Switzerland

Phone: +41 (0)22 917 44 44
Fax: +41 (0)22 917 05 05
E-mail: info.ece@unece.org
Website: http://www.unece.org

The mission for the project took place from 9 to 21 June 2003.

LIST OF CONTRIBUTORS

Azerbaijan

Mr. Cussein Bagirov Minister	Ministry of Ecology and Natural Resources
Mr. Issa Aliyev	Ministry of Ecology and Natural Resources
Mr. Rasim Sattarzade	Ministry of Ecology and Natural Resources
Mr. Emin Garabaghli	Ministry of Ecology and Natural Resources
Mr. Ali Orujov Abdulali	Ministry of Ecology and Natural Resources
Mr. Mutallim Abdulhasanov	Ministry of Ecology and Natural Resources
Mr. Imran Abdulov	Ministry of Ecology and Natural Resources
Ms. Maisa Adigezalova	Ministry of Ecology and Natural Resources
Mr. Lahuti Aliyev	Ministry of Ecology and Natural Resources
Mr. Magsud Babayev	Ministry of Ecology and Natural Resources
Mr. Mamed Djafarov	Ministry of Ecology and Natural Resources
Ms. Saijad Efendieva	Ministry of Ecology and Natural Resources
Mr. Farig Farzaliyev	Ministry of Ecology and Natural Resources
Mr. Baghir Hidayatov	Ministry of Ecology and Natural Resources
Mr. Oktay Jafarov	Ministry of Ecology and Natural Resources
Mr. Gussein Mamedov	Ministry of Ecology and Natural Resources
Ms. Sadagat Mamedova	Ministry of Ecology and Natural Resources
Mr. Gabil Mammadov	Ministry of Ecology and Natural Resource
Mr. Huseyn Mammadov	Ministry of Ecology and Natural Resource
Mr. Mammadhusein Muslimov	Ministry of Ecology and Natural Resources
Mr. Mamed Nasibov	Ministry of Ecology and Natural Resources
Mr. Ali Saidov	Ministry of Ecology and Natural Resources
Mr. Mussa Shakiliev	Ministry of Ecology and Natural Resources
Mr. Elshan Asadov	Ministry of Economic Development
Mr. Rustam Makhmudov	Ministry of Economic Development
Mr. Mehman Rzayev	Ministry of Economic Development
Mr. Fizuli Akhmedov	Ministry of Education
Mr. Nuradin Abdullayev	Ministry of Health
Ms. Samaya Ahmedova	Ministry of Health
Mr. Oktay Akhundov	Ministry of Health
Mr. Fikret Aslanov	Ministry of Health
Mr. Victor Gasimov	Ministry of Health
Mr. Viktor Gasimov	Ministry of Health
Mr. Azer Maharrmow	Ministry of Health
Mr. Fuad Mardanli	Ministry of Health
Mr. Aliev Namiq Neriman	Ministry of Health
Ms. Leyla Tagizadeh	Ministry of Health
Mr. Abbass Valibayov	Ministry of Health
Mr. Natig Amirov	Ministry of Taxes
Ms. Rana Lazymova	State Statistical Committee

Ms. Basti Bagirova	Azerbaijan Medical University
Mr. Emil Hasanov	Azerbaijan National Agency for Mine Action (ANAMA)
Ms. Nazim Ismaylov	Azerbaijan National Agency for Mine Action (ANAMA)
Mr. Eldar Abdullayev	Center for Hygiene and Epidemiology of Baku City
Mr. Nadir G. Eyvazov	Republican Hygienic and Epidemiological Center
Mr. Rza Gasimov	Republic Centre for Hygiene and Epidemiology
Mr. Sullayman Mamedov	Republic Centre for Hygiene and Epidemiology
Mr. Murad Mamedov	Republican Oncological Institute
Mr. Shamil Shikhaliev	
Mr. Agamakhmud Sirajov	State Oil Company of the Republic of Azerbaijan

Nongovernmental organizations

Mr. Talat Kangarli	AZE Republic Public Association for Sustainable Development Assistance (CHEVRA)
Mr. Samir Isaev	Ecolex Azerbaijan Environmental Law Centre

TABLE OF CONTENTS

LIST OF FIGURES

LIST OF TABLES

LIST OF BOXES

ACRONYMS AND ABBREVIATIONS

ADB	Asian Development Bank
ARRLA	Agency for Rehabilitation of Reconstruction of Liberated Areas
ARWC	Absheron Regional Water Company
AZM	Azerbaijan National Currency (Manat)
BAT	Best Available Technique
BOD	Biological oxygen demand
CENN	Caucasus environmental NGO network
CFCs	Chlorofluorocarbons
CIDA	Canadian International Development Agency
CITES	Convention on International Trade in Endangered Species of Wild Fauna and Flora
CMNAR	Cabinet of Ministers of Nakhchevan Autonomous Republic
CMRA	Cabinet of Ministers of the Republic of Azerbaijan
COD	Chemical oxygen demand
CPI	Consumer price index
DDT	Dichlorodiphenyltrichloroethane
EAP TF	OECD Task Force for the Implementation of the Environmental Action Programme for Central And Eastern Europe
EBRD	European Bank for Reconstruction and Development
EC	European Commission
EEA	European Environment Agency
EECCA	Eastern Europe, Caucasus and Central Asia
EIA	Environmental Impact Assessment
EIB	European Investment Bank
EME	Environmental Management in Enterprises
EMEP	Cooperative Programme for Monitoring and Evaluation of the Long-rangeTransmission of Air Pollutants in Europe
EMS	Environmental Management System
EPR	Environmental Performance Review
EQO	Environmental Quality Objective
EQS	Environmental Quality Standard
EU	European Union
EU TACIS	European Union Technical Assistance to Commonwealth of Independent States
FAO	Food and Agriculture Organization of the United Nations
FDI	Foreign direct investment
GDP	Gross domestic product
GEF	Global Environmental Facility
GHGs	Greenhouse gases
GIS	Geographic information system
GTZ	German Agency for Technical Co-operation
HCFCs	Hydrochlorofluorocarbons
HDI	Human Development Index
HESME	Health, Environment and Safety Management in Enterprises
HMs	Heavy metals
HPP	Hydropower plant
IAEA	International Atomic Energy Agency
IDB	Islamic Development Bank
IDPs	Internally Displaced Persons
IFI	International financing institution
ILO	International Labour Organisation
IMF	International Monetary Fund
IPCC	Intergovernmental panel on Climate Change
IPPC	Integrated pollution prevention and control
ISO	International Standardization Organization

IUCN	World Conservation Union
JBIC	Japanese Bank of International Cooperation
JICA	Japan International Cooperation Agency
LEAPs	Local Environmental Action Plans
MAB	Man and Biosphere Programme of UNESCO
MAC	Maximum allowable concentration
MAD	Maximum allowable discharge
MEAs	Multilateral environmental agreements
NATO	North Atlantic Treaty Organisation
NB	National Bank of Azerbaijan
NEAP	National Environmental Action Plan (Program)
NEHAP	National Environment and Health Action Plan
NGOs	Non–governmental Organizations
NMVOC	Non-methane volatile organic compounds
NPESSED	National Programme on Ecologically Sustainable Socio-Economic Development
ODS	Ozone-depleting substances
OECD	Organisation for Economic Co-operation and Development
OSCE	Organization for Security and Co-operation in Europe
PAH	Polyaromatic hydrocarbon
PCA	Partnership and Co-operation Agreement
PCB	Polychlorinated biphenyl
PIC	Prior informed consent
POPs	Persistent organic pollutants
PPI	Producer price index
PPP	Purchasing power parity
PRTR	Pollution Release and Transfer Registry
REC	Regional Environmental Center for the Caucasus
REReP	Regional Environmental Reconstruction Programme
SEA	Strategic environmental assessment
SME	Small and Medium Enterprises
SOCAR	any of Azerbaijan Republic
SOFAR	State Oil Fund of Azerbaijan Republic
SPM	Suspended particulate matter
SPPRED	State Programme on Poverty Reduction and Economic Development
UEIP	Urgent Environmental Investment Projects
UNDP	UN Development Programme
UNECE	United Nations Economic Commission for Europe
UNEP	United Nations Environment Programme
UNFPA	United Nations Population Fund
UNHCR	UN High Commissioner for Refugees
UNICEF	United Nations Children's Fund
UNIDO	United Nations Industrial Development Organization
USAID	United States Agency for International Development
USD	US Dollars
VAT	Value-added tax
VOCs	Volatile organic compounds
WB	World Bank
WHO	World Health Organisation
WWF	World Wide Fund for the Nature

SIGNS AND MEASURES

..	not available
-	nil or negligible
.	decimal point
ha	hectare
kt	kiloton
g	gram
kg	kilogram
mg	milligram
mm	millimetre
cm^2	square centimetre
m^3	cubic metre
km	kilometre
km^2	square kilometre
toe	ton oil equivalent
l	litre
ml	millilitre
min	minute
s	second
m	metre
°C	degree Celsius
GJ	gigajoule
kW_{el}	kilowatt (electric)
KWh	kilowatt-hour
kW_{th}	kilowatt (thermal)
MW_{el}	megawatt (electric)
MW_{th}	megawatt (thermal)
MWh	megawatt-hour
GWh	gigawatt-hour
TWh	terawatt-hour
Bq	becquerel
Ci	curie
MSv	millisievert
Cap	capita
Eq	equivalent
H	hour
kv	kilovolt
MW	megawatt
Gcal	gigacalorie
Hz	hertz

Currency

Monetary unit: Azerbaijan Manat = 100 kopeks

Exchange rates: IMF

Year	Manat/US$	Manat/Euro
1995	4,413.54	5,773.35
1996	4,301.26	5,454.00
1997	3,985.38	4,519.82
1998	3,869.00	4,333.28
1999	4,120.17	4,395.40
2000	4,474.15	4,134.11
2001	4,656.58	4,170.43
2002	4,860.82	4,590.56

Source: IMF. International Financial Statistics, September 2003.
Note: Values are annual averages. NC/Euro rate is calculated using US$/Euro rates from the IMF IFS September 2003.

INTRODUCTION

I.1   The physical context

Azerbaijan, the largest of the three republics of the South Caucasus, occupies the southern part of the isthmus between the Black Sea and the Caspian Sea. The country has a total land area of 86,600 km^2 and it is bordered by the Russian Federation to the north, the Caspian Sea to the east, the Islamic Republic of Iran to the south, Armenia to the west, Georgia to the north-west and Turkey to the south-west. The Nakhchivan Autonomous Republic of Azerbaijan is separated from the rest of Azerbaijan by a strip of Armenian territory. The national frontier is 2013 km long.

The borders of Azerbaijan generally correspond to natural geographic features. The western coast of the Caspian Sea forms the country's entire eastern border. The Greater Caucasus mountain range forms part of Azerbaijan's northern border with the Russian Federation and contains the country's highest peak, Mount Bazardüzü (4,466 m). The Greater Caucasus extends into north-east Azerbaijan and runs south-east as far as the Absheron Peninsula, which protrudes into the Caspian Sea. The Lesser Caucasus range in western Azerbaijan is smaller, reaching only to about 3,500 m and forming part of the border with Armenia. The extreme south-east border of Azerbaijan is formed by the Talish mountains.

Azerbaijan's two main rivers, the Kura (1,515 km long) and its tributary the Araz (1,072 km long), both rise in the mountains of north-east Turkey. The Kura flows into north-west Azerbaijan from neighbouring Georgia and then follows a south-easterly course to the Caspian Sea. The Araz forms part of Azerbaijan's southern border with the Islamic Republic of Iran and turns north-east to enter south-central Azerbaijan, where it joins the Kura in Sabirabad and drains into the Caspian Sea.

Azerbaijan has a range of climates: cold weather in the Great Caucasus mountain range with heavy snowfall during the winter above 3,000 m; a temperate climate in the Kura plain; and a subtropical climate in the Lenkeran lowlands (which receive significantly more precipitation than other areas of the country), in the south-east on the Caspian coast. In the lowlands winters are relatively mild with high precipitation, and summers are hot, dry and long. The average lowland temperature in July is +27°C, sometimes reaching the upper 30s. In January, the average temperature there is +1°C. The capital Baku has a moderately warm and dry subtropical climate, with a hot summer and a short, mild winter. Frosts occur only once every 10 years. Baku is susceptible to year-round strong winds; it has the least rainfall and the greatest annual average number (284) of sunny days in the Caucasus.

Azerbaijan has a variety of mineral resources including iron ore, aluminium, copper, lead, zinc, limestone and salt, but its most important raw materials are crude oil and gas. Most of Azerbaijan's oil reserves are located offshore, beneath the Caspian Sea, with most developed oilfields near the Absheron Peninsula. Although Azerbaijan produces natural gas and has large potential reserves, it is a net importer. The infrastructure is insufficient to deliver the country's natural gas from offshore fields to the mainland, and therefore instead of being piped to markets natural gas is flared off.

Azerbaijan's power sector has an installed generating capacity of approximately 5.1 gigawatts (GW). Electricity generation totalled 18.2 terawatt-hours (TWh) in 2001, and consumption 16.6 TWh. However, Azerbaijan is a net electricity importer because much of its power generation is lost in transmission. Minimal public investment and maintenance since independence have left the power infrastructure in a generally poor condition.

Figure I.1: Land use, 1998

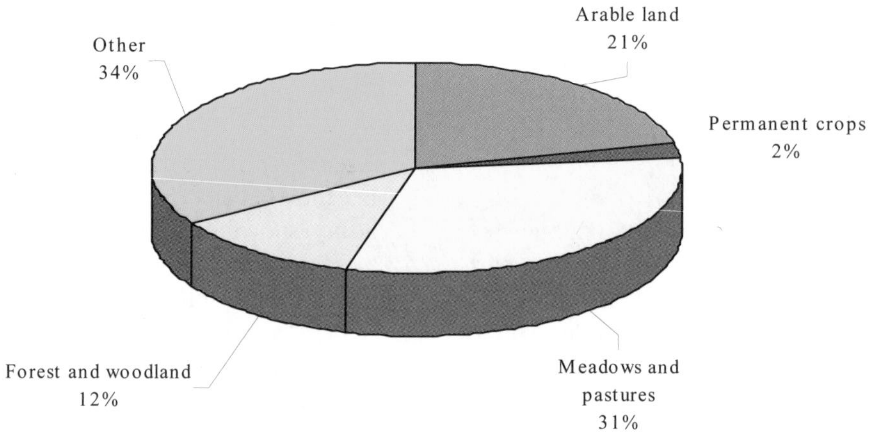

Source: The State Statistical Committee. Food Security in Azerbaijan 2001 (Statistical Yearbook).

The agricultural sector produced 16.8% of GDP in 2001 compared to 26.7% in 1995. (see figure I.2) It still employed 40% of the labour force in 2001. The primary food crops produced are barley, maize, potatoes, rice, soybeans, sugar beet and wheat. The primary meat products are beef and veal, chicken, lamb and pork. While cotton is the leading cash crop, the production of wine grapes and tobacco is also significant. In 1998 the largest agricultural exports in value were cotton, tobacco, wine, alcoholic beverages, and cottonseed oil. (See figure I.1 and chapter 9 on land use, agriculture and desertification.)

I.2   The human context

Azerbaijan has the largest population of the states of the South Caucasus, but it is also the least urbanized, with 51% of the population living in urban areas. The total population of Azerbaijan in 2001 was 8.1 million inhabitants, with an average population density of 95 persons/km². The most densely populated area is the Absheron Peninsula in the east, where the capital Baku (pop. 1,828,800) is located. Other industrial centres and important towns are Ganja (301,400) in the west and Sumgayit (288,400) on the Caspian coast.

The ethnic composition of Azerbaijan changed during the decade following 1989. In 1989, 80% of the population was *Azerbaijani,* but in 1999, 90.6% of the population was Azerbaijani. Minorities in Azerbaijan include the Lezgins (2.2%), the Russians (1.8%), the Armenians (1.5%), the Talish

(1%) and ten other population groups who represent less than 1% of the population.

The demographic indicators of Azerbaijan have changed markedly during the past ten years. The fertility rate decreased from 2.8 in 1991 to 1.9 in 2001. The birth rate also fell sharply from 26.1 (per 1000) in 1990 to 13.6 in 2001. The infant mortality rate dropped from 22.9 (per 1000) in 1990 to 12.5 in 2001. The average life expectancy was 72.4 years (75 years for women and 69.7 years for men) in 2001. (See table I.1).

During the past few years the country's human development has been positive. In 1997, Azerbaijan's Human Development Index (HDI) as calculated by the United Nations Development Programme (UNDP) stood at 0.695 (on a scale of 0.0 to 1.0) and the country ranked 103rd out of 174 countries. Its HDI has improved steadily, and in 2001 it was the 89th country out of 175 with an HDI of 0.744.

The official language of the country is Azeri, a Turkic language, closely related to Turkish and Turkmen. Originally Azeri was written with Arabic script, but in the early 1920s Latin script was introduced. Under the Soviet Union, it was changed to Cyrillic, but this was abandoned after independence and replaced with the current Turkish version of Latin script. The country's literacy rate in 1999 was 97% and the attainment quotient of post-secondary education for adults aged 25 and older was 14% in 1991.

Figure I.2: GDP composition by sector: 1995 and 2001 (percent of the total GDP

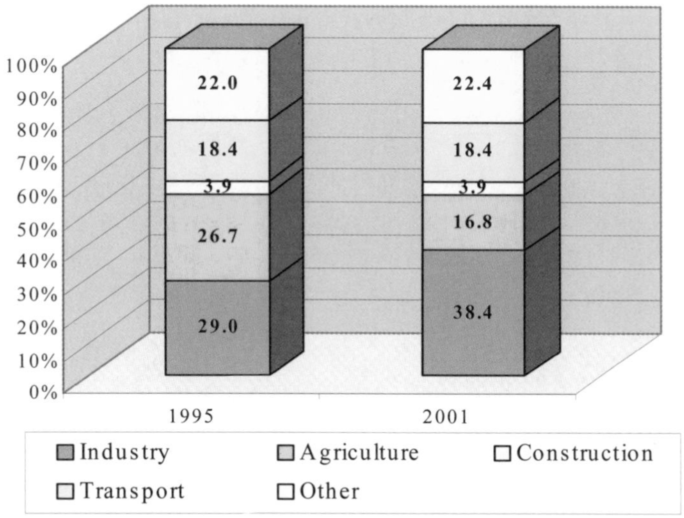

Source: UNECE common statistical database, 2003.

The traditional religion of Azerbaijan is Islam. Most Azerbaijanis identify themselves as Muslims. The majority of Muslims (about 70%) are Shi'ah and the rest are Sunni. Religious minorities include Russian Orthodox, Armenian Orthodox and other smaller groups.

I.3 The political and economic context

Political context

In 1988-1990 the national democratic movement in Azerbaijan campaigned for the restoration of the country's independence. On 23 September 1989, Azerbaijan was among the first Soviet republics to adopt its own Constitutional Law on Sovereignty. The struggle for independence culminated in the adoption on 31 August 1991, by the Supreme Council of the Republic of Azerbaijan, of a declaration on the restoration of the State independence of the Republic of Azerbaijan. This was followed on 18 October of the same year with the passage of the act establishing the State independence of the Republic of Azerbaijan, which set out the foundations for the statehood of an independent Azerbaijan and determined the principles of its political and economic structure.

Throughout this period, the internal politics of the country was largely dominated by the tensions between Armenia and Azerbaijan concerning the conflict in and around the

Nagorny Karabakh region of Azerbaijan, which resulted in occupation of 20% of the territory of Azerbaijan. The United Nations estimated that nearly 1 million refugees and internally displaced persons were in Azerbaijan at the end of 1993. Four United Nations Security Council Resolutions (822, 853, 874, 884) concerning the conflict were adopted in 1993. These Resolutions reaffirmed the sovereignty and territorial integrity of Azerbaijan, demanded immediate cessation of all hostilities and the immediate, complete and unconditional withdrawal of all occupying forces from the territories of Azerbaijan. They also called for creation of all necessary conditions to allow the refugees and displaced persons to return to their homes in security and dignity.

Economic context

Oil has been the driving force of Azerbaijan's economy since Baku emerged as one of the earliest centres of the petroleum industry more than 100 years ago. The country experienced an oil boom at the beginning of the 20th century and later served as a major refining and oilfield equipment production centre for the former Soviet Union. Oil production peaked at about 500,000 barrels per day (bbl/day) during the Second World War, but fell significantly after the 1950s as the Soviet Union redirected resources elsewhere.

Table I.1: Demography and health indices, 1990-2001

	1990	1991	1992	1993	1994	1995	1996	1997	1998	1999	2000	2001
Birth rate (per 1000)	26.1	26.5	25.0	23.8	21.6	19.3	17.3	17.5	15.7	14.7	14.5	13.6
Fertility rate	..	2.8	2.3	2.5	2.7	2.5	..	2.7	2.0	2.0	1.9	1.9
Mortality rate (per 1000)	6.1	6.2	7.1	7.2	7.4	6.8	6.4	6.2	5.9	5.8	5.8	5.6
Infant mortality rate (per 1000)	22.9	25.0	26.0	28.6	26.2	24.3	20.8	19.4	16.6	16.5	12.8	12.5
Female life expectancy at birth (years)	75.3	75.2	73.7	73.3	73.1	73.4	74.3	74.8	74.8	74.1	74.4	75.0
Male life expectancy at birth (years)	67.1	66.6	64.1	63.9	62.9	65.4	66.4	67.0	68.1	68.3	68.7	69.7
Life expectancy at birth (years)	71.4	71.1	68.9	68.6	67.9	69.5	70.4	71.0	71.5	71.2	71.6	72.4
Population aged 1-14 in total (%)	33.2	33.3	33.4	33.4	33.3	33.0	32.8	32.8	31.7	31.3	30.3	29.2
Population aged 65 or over in total (%)	4.8	4.8	4.9	5.1	5.4	5.6	5.7	5.7	5.6	5.6	5.8	6.1

Sources : WHO. Health for All database. www.who.dk on 21.8.2003.

Azerbaijan passed its first law on the privatization of State property in 1993, but privatization was delayed due to the conflict in Nagorny-Karabakh. In the first phase of privatization, from 1996 to 1998, about 21,000 small and 1,000 medium- and large-size enterprises were privatized. Virtually all small businesses and 90% of agriculture have been privatized.

Azerbaijan is struggling to overcome the economic collapse that followed independence. In 1995 its GDP was only 37.0% of its 1989 level. By 2002, economic recovery had raised GDP to 63.1% of the 1989 level. Independence in 1991 also brought along changes in GDP composition. The industrial sector's share rose from 24.9% in 1991 to 38.4% in 2001. Agriculture's share fell from 32.3% in 1991 to 16.8% in 2001. Azerbaijan's economy is still much smaller than it was in 1991; even with the expected strong growth, the economy is not expected to return to its 1991 level until about 2007.

In 2000 the growth of oil production and exports resulted in the first trade surplus since 1992. The value of the export of goods approximately doubled from about US$1 billion in 1999 to over US$2 billion in 2001. While the current account remains in deficit, this deficit is diminishing fast: in 2001 it was US$ 51.7 million (about 1% of GDP) as compared to US$ 167.7 million in 2000 (3% of GDP).

The capital and financial account is dominated by foreign direct investment in the oil and gas sector. The overall balance of payment surpluses in recent years has been large enough that Azerbaijan was able to build international reserves

of nearly US$ 900 million by the year-end of 2001, a secure seven months' worth of imports.

Azerbaijan's future economic growth relies in large part on the successful development of its oil and natural gas resources. Crude oil and oil product exports make up over 70% of exports, and oil-related revenue makes up nearly 50% of budget revenue. Some industry experts estimate that Azerbaijan could be exporting 1 million bbl/day by 2010. (See chapter 11, environmental concerns in oil and gas sectors.)

I.4 The institutions

Azerbaijan is a democratic, secular, unitary republic. The current constitution of Azerbaijan was adopted by referendum in 1995 and amended by the referendum held on 24 August 2002.

Azerbaijan is a presidential republic. The President, directly elected for not more than two consecutive five-year terms, is the head of State. The President appoints the Prime Minister and forms the Government. The current president is Ilham Aliyev, who was elected on 15 October 2003, with 76.84% of the votes.

State power is based on the principle of division of powers: legislative power is implemented by the Parliament, executive power by President and the judiciary, through the courts.

The unicameral National Assembly, Milli Mejlis, has 125 members. The winners of the district elections fill 99 seats and 25 are awarded according to the proportion of votes that the parties have gained in the elections on the national level. The remaining seat for Nagorny-Karabakh is kept vacant. Parliamentary elections are held every five years. The President can veto the decision of the Parliament, which needs a majority vote (95 votes) to override the presidential veto. The judicial branch has a Supreme Court, an Economic Court and a Constitutional Court. The President nominates their judges. (See chapter 1, policy, legal and institutional framework.)

I.5 Environmental context:

The key environmental issues and problems identified in the National Environmental Action Plan are:

- Severe air, water and soil pollution from industry, oil production, energy and transport, particularly in the Absheron Peninsula, including Baku and Sumgayit, and the Caspian Sea;
- Oil-contaminated sites and abandoned industrial facilities;
- Environmentally unsound disposal of municipal and industrial waste;
- Threat of irreversible loss of the Sturgeon stock;
- Limited water resources and high losses during distribution;
- The Caspian Sea water level fluctuation causing extensive flooding along Azerbaijan's coast since 1978;
- Degradation of agricultural land, loss of forestry and biodiversity;
- Loss of cultural heritage. Many of Azerbaijan's historical sites are in a serious state of disrepair or deterioration. Some of its architectural monuments are structurally damaged and unsound.

These issues are discussed in depth in the chapters that follow.

Table I.2: Selected economic indicators, 1990-2001

	1990	1991	1992	1993	1994	1995	1996	1997	1998	1999	2000	2001
GDP (change, 1989=100)	88.3	87.7	67.9	52.2	41.9	37.0	37.4	39.6	43.6	46.8	52.0	57.1
GDP (% change over previous year)	-11.7	-0.7	-22.6	-23.1	-19.7	-11.8	1.3	5.8	10.0	7.4	11.1	9.9
GDP in current prices (million manat)	1,466	2,672	24,101	157,082	1,873,387	10,669,000	13,663,200	15,791,400	17,203,100	18,875,400	23,590,500	26,578,000
GDP in current prices (million US$)	..	..	..	..	1,200.0	2,417.3	3,176.6	3,962.3	4,446.3	4,581.2	5,272.7	5,657.0
GDP per capita (US$ PPP per capita)	4,123.4	4,187.3	3,270.1	2,536.3	2,052.3	1,828.1	1,867.8	1,995.4	2,200.4	2,374.5	2,676.2	2,982.2
Share of agriculture and fishing in GDP (%)	29.0	32.3	27.7	26.7	31.4	26.7	27.2	21.5	18.6	19.1	17.0	16.8
Industrial output (% change over previous year)	-6.3	-8.9	-30.4	-19.7	-24.7	-21.4	-6.7	0.3	2.2	3.6	6.9	5.1
Agricultural output (% change over previous year)	1.0	0.3	-23.8	-15.5	-12.8	-7.0	3.0	-6.1	6.2	7.1	12.1	11.1
Labour productivity in industry (% change over previous year)	-3.5	-6.8	-25.7	-12.0	-21.1	-16.6	16.1	17.3	-1.5	0.4	11.0	6.1
CPI (% change over the preceding year, annual average)	6.1	106.6	912.3	1,129.1	1,663.5	411.8	19.9	3.7	-0.8	-8.5	1.8	1.5
PPI (% change over the preceding year, annual average)	..	179.5	7,453.6	1,974.0	3,779.0	384.9	67.7	11.4	-12.4	-9.3	24.6	1.8
Registered unemployment (% of labour force, end of period)	..	..	0.2	0.7	0.9	1.1	1.1	1.3	1.4	1.2	1.2	1.3
Balance of trade in goods and non-factor services (million US$)	..	..	489.0	-122.0	-163.0	-373.1	-693.9	-566.9	-1,046.1	-408.2	319.4	613.9
Current account balance (million US$)	..	153.0	488.0	-160.0	-121.0	-400.7	-931.2	-915.8	-1,364.5	-599.7	-167.7	-51.7
" (as % of GDP)	..	..	..	..	-10.1	-16.6	-29.3	-23.1	-30.7	-13.1	-3.2	-0.9
Net FDI inflows (million US$)	..	..	0.0	60.0	22.0	330.0	627.3	1,124.8	1,022.9	510.4	129.1	226.5
Net FDI flows (as % of GDP)	..	..	..	..	1.8	13.7	19.7	28.4	23.0	11.1	2.4	4.0
Cumulative FDI (million US$)	0.0	0.0	0.0	60.0	82.0	412.0	1,039.3	2,164.1	3,187.0	3,697.4	3,826.5	4,053.0
Foreign exchange reserves (million US$)	..	1.0	1.0	1.0	2.0	120.9	211.3	466.1	447.3	672.6	680.0	896.7
(as months of imports)	..	0.04	0.02	0.01	0.03	1.47	1.90	4.07	3.11	5.63	5.30	7.34
Total net external debt (million US$)	..	1.0	1.0	52.0	240.0	418.0	522.0	452.0	665.0	968.0	1,170.0	1,250.0
Exports of goods (million US$)	..	395.0	1,275.0	697.0	682.0	612.3	643.7	808.3	677.8	1,025.2	1,858.3	2,078.9
Imports of goods (million US$)	..	336.0	786.0	819.0	845.0	985.4	1,337.6	1,375.2	1,723.9	1,433.4	1,538.9	1,465.0
Ratio of net debt to exports (%) (calc)	..	0.3	0.1	7.5	35.2	68.3	81.1	55.9	98.1	94.4	63.0	60.1
Ratio of gross debt to GDP (%)	..	..	..	..	20.0	17.3	16.4	11.4	15.0	21.1	22.2	22.1
Exchange rates: annual averages (manat/US$)	..	..	..	99.98	1,570.2	4,413.5	4,301.3	3,985.4	3,869.0	4,120.2	4,474.2	4,656.6
Population (1000)	7,174.9	7,271.3	7,382.1	7,494.8	7,596.6	7,684.9	7,763.0	7,838.3	7,912.9	7,982.6	8,048.6	8,111.2

Source: UNECE Common statistical database and National Statistics, 2003.

Table I.3: List of Ministries and related organizations in Azerbaijan

Cabinet of Ministers
Ministry of Agriculture
Ministry of Communication
Ministry of Culture
Ministry of Defence
Ministry of Ecology and Natural Resources
Ministry of Ecology and Natural Resources of Nakhchivan autonomous republic
Ministry of Economic Development
Ministry of Education
Ministry of Finances
Ministry of Foreign Affairs
Ministry of Fuel and Energy
Ministry of Health
Ministry of Internal Affairs
Ministry of Justice
Ministry of Labour and Protection of the Population
Ministry of National Security
Ministry of Taxes
Ministry of Youth, Sport and Tourism
National Department of Environmental Monitoring (NDEM) - under MENR
State Commission on Climate Change
State Committee for Biodiversity and Genetic Resources
State Committee for Land and Cartography
State Committee for Standardization - State Agency for Standardization, Metrology and Patents
State Committee of Amelioration and Water Management - Committee for Amelioration and Water
Farm attached to the Cabinet of Ministers
State Committee of Architecture and Construction
State Control Inspectorate for Environment and Use of Natural Resources (SCI) under MENR
State Hydrometeorological Services - under MENR
State Sanitary Epidemiological Service - under the Ministry of Health
State Statistical Committee

Source: Ministry of Ecology and Natural Resources, 2003.

Figure I.3: Map of Azerbaijan

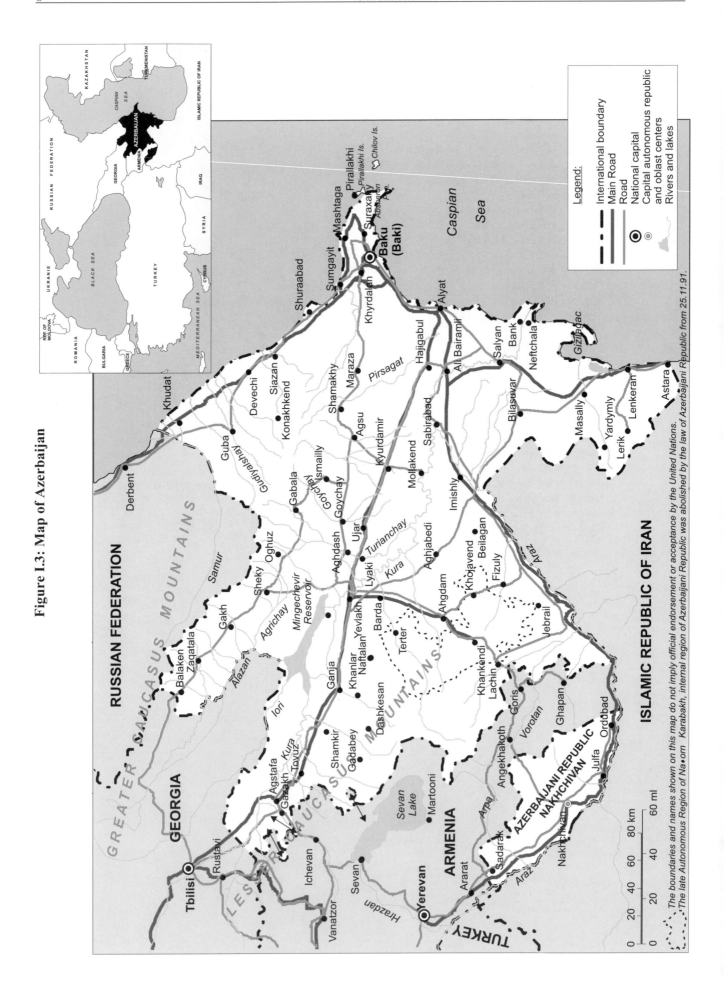

Legend:
International boundary
Main Road
Road
National capital
Capital autonomous republic and oblast centers
Rivers and lakes

PART I: THE FRAMEWORK FOR ENVIRONMENTAL POLICY AND MANAGEMENT

Chapter 1

POLICY, LEGAL AND INSTITUTIONAL FRAMEWORK

1.1 Introduction

During its 12 years of independence, Azerbaijan has steadily improved its system of environmental protection. The policy, legal and institutional framework that it inherited from the former Soviet Union was not designed to operate within a market economy, and insufficient attention had been given to issues of efficiency and environmental protection.

Much progress has now been made, particularly in updating the environmental legal framework, although further improvements are still needed, including in environmental impact assessment. The new Ministry of Ecology and Natural Resources was established in 2001 and other institutional reform is being undertaken. Particular attention needs to be given in this regard to the organization and effectiveness of the inspectorates. Finally, a number of good policies for environment, poverty and sustainable development have been developed, but their relationship remains to be clarified.

1.2 Policy Framework

National Environmental Action Plan (NEAP)

The first National Environmental Action Plan of Azerbaijan was developed from 1995 to 1998 with World Bank assistance, in parallel with the first "wave" of NEAPs in Eastern Europe, the Caucasus and Central Asia. Based on a description of the current situation, it identifies the following problems requiring urgent actions:

- Severe pollution caused by industries, oil exploration and production, and energy;
- Threat of irreversible collapse of the sturgeon stock triggered by a loss of reproductive capacity, pollution, and overfishing;
- Deteriorating water quality, especially of drinking water, both in rural and in urban areas, causing an increase in water-borne diseases;
- Loss of fertile agricultural land from erosion, salinization, pollution with heavy metals and chemicals, and deteriorating irrigation systems;

- Threats to protected areas leading to losses in biodiversity;
- Loss of forestry cover, mainly in war-affected areas;
- Damage to the Caspian coastal zone caused by flooding from a rise in sea level and pollution;
- Deterioration of cultural heritage, due to natural causes, aggravated by modern environmental problems such as acid rain and uncontrolled development.

The NEAP also stresses the importance of policy reform and of integrating environmental and economic policies. Stemming from the description and analysis of those problems, the NEAP puts forward a list of environmental priorities, setting 32 objectives grouped in five categories:

- Pollution from industrial production, energy production, transport and other sources;
- The Caspian Sea;
- Forestry, land and biodiversity;
- Institutional development; and
- Policy.

Actions to achieve the objectives are identified and prioritized on the basis of human health, irreversible damage to natural resources and impediments to economic development. All actions are assigned to a specific government agency responsible for their implementation.

According to the Plan, the cost of each action varies between US$ 5,000 and 5,000,000, and totals US$ 42.5 million. Actions identified as top priority are to be completed within one to two years; other priorities within two to five years. Of the 46 actions, 33 are top priority.

If fully financed by the State budget, the NEAP would have required just above 1% of State budget expenditures if completed within the five-year time frame (1998-2003) and up to 3% if accomplished mostly within one to two years. This was unrealistic if perceived as a wholly environmental expenditure (given the traditionally low priority of environmental protection in Eastern Europe, the Caucasus and Central Asia). Funds would need to be provided either from other budget lines or from other sources. Indeed, the NEAP notes that "financing should be sought from a combination of the Government of Azerbaijan, international donors and private enterprises", but is not more specific.

It is therefore not surprising that, by the end of the five-year period, only a small percentage of the activities (around 20%) listed in the NEAP have been accomplished. In general, there has been insufficient environmental progress in the economic sectors, including in the prevention of new pollution and the clean-up of old pollution. Some of the NEAP policy elements have been achieved, particularly with regard to new legislation, but many of the implementing regulations and by-laws are lacking. Small-scale developments have taken place in forestry and biodiversity. The most significant change was institutional, as Azerbaijan established a Ministry of Ecology and Natural Resources for the first time.

Other than financing, a significant obstacle to implementation has been the lack of more specific prioritization. The current categories of "top priority" and "priority" do not provide sufficient guidance for the distribution of limited resources.

In addition, it appears that only the direct costs of proposed actions were taken into consideration. For instance, the action to "sign and ratify key international conventions" was said to require just US$ 5,000. No consideration was given to "entailed" costs of compliance to be borne by the State budget or others, e.g. enterprises. This would tend to suggest that there had been no cost-benefit analysis.

The limitations mentioned above are typical for the first generation of NEAPs in Eastern Europe, the Caucasus and Central Asia. Even with these limitations, the first NEAP of Azerbaijan has had a very positive effect on the development of environmental and natural resource protection, thus proving the value of NEAP as a policy instrument. This impact is mainly twofold:

- The NEAP has for the first time secured a broad consensus among stakeholders (government, civil society and international partners) concerning the urgency of the measures to protect the national environment, thus setting a common basis for future actions. Several NEAP projects have attracted international funding;

- The NEAP influenced institutional changes in the government and triggered the establishment of more consolidated and powerful environmental institutions – including, first and foremost, the Ministry of Ecology and Natural Resources.

State Programme on Poverty Reduction and Economic Development for 2003-2005

Approved by presidential decree on 20 February 2003, this document is envisaged to play a significant role in the medium term; as a comprehensive strategy with a multi-sector approach, it will influence the environment sector within the context of overall national priorities.

In accordance with presidential decree 636 of 2 March 2001, a State commission, led by the Prime Minister, was established to develop the Programme. Fifteen sectoral working groups, with members from government institutions, non-governmental organizations and civil society, were established to prepare the final document. A secretariat, supported by foreign experts, was established at the Ministry of Economic Development.

The Programme covers a period of three years (2003-2005) and will be revised annually as the envisaged policy measures are implemented and yield results. Among other things, it addresses the role of environmental conditions as a cause of poverty as well as a tool to reduce it. Based on the assumption that "economic development which upsets the environmental balance cannot be sustainable", it obliges the Government to promote balanced growth and to bring about improvements in some of the key economic sectors: to improve the investment climate, to increase access to credit among businesses and entrepreneurs, to develop the infrastructure, to encourage small and medium enterprises, to develop the regions and agriculture, to improve the environment, to reform energy generation and distribution, and to promote tourism.

While enumerating environmental problems (water, land, air, forests, Caspian Sea), the Programme uses language close to that of the NEAP. Some of the solutions suggested in the Programme are:

- Tackling large-scale national and international projects, such as those concerning the Caspian Sea;

- Improving the management of the country's natural resources;
- Improving the regular monitoring of environmental pollution;
- Conducting campaigns to raise public awareness of the need to respect the ecological balances.

In particular, the Programme recommends addressing the following with public finances (again consistent with the NEAP):

- Clean up mercury sludge and improve waste management;
- Reconstruct the water purification units and sewerage system in Sumgayit, expand the purification facilities in Baku, and finish construction of new facilities in Ganja;
- Protect the traditional spawning areas for sturgeon in the Kura and Araz rivers;
- Treat the contaminated water discharged from the Zykh and other lakes into the Caspian Sea;
- Reclaim oil-polluted soil in the Absheron Peninsula;
- Treat radioactive waste in the coal storages at Ramany, New-Surakhany and Neftchala iodine-processing plants;
- Reverse negative trends in forestry by curbing uncontrolled logging and replanting (some 15,500 ha in total);
- In the long term, develop alternative energy sources, including geothermal energy sources, and a national programme for solar and wind power generation;
- Enforce environmental protection through the improvement of monitoring of environmental indicators;
- Carry out administrative reforms to improve management;
- Provide regional laboratories with the necessary equipment for monitoring;
- Involve municipalities and communities in the assessment of the environmental impact of regional economic development projects; and
- Improve waste collection and disposal, and rehabilitate sewage systems at municipal levels, taking advantage of best available technologies at affordable costs to introduce recycling practices and reduce water and soil contamination.

The Programme, however, fails to set priorities, nor does it provide an assessment of costs and benefits. Logically, the agencies responsible (including the Ministry of Ecology and Natural Resources) should complete the Programme by developing concrete projects and plans for each of the objectives and rank them on the basis of common criteria; yet the Programme does not give any instruction or reference to the methodology to be applied.

National Programme on Environmentally Sustainable Socio-economic Development (2003)

The Ministry of Ecology and Natural Resources prepared the National Programme on Environmentally Sustainable Socio-economic Development as requested in presidential decree 612 (para. 19), on the Implementation of the Law on the State Budget for 2002 (26 December 2001). It was endorsed by presidential decree 1152, Approving the National Programmes on Ecology (18 February 2003).

The Programme covers the environmental aspects of the country's overall development strategy. It determines the main areas of sustainable development and includes a plan of action for 2003-2010 "to address the initial phase of the resolution of the current problems". The Ministry of Ecology and Natural Resources is to provide all necessary guidelines, application software and scientific data for its full implementation.

Attached to the Programme is an action plan for its implementation, identifying the actions, the main implementing agencies and the timelines for a range of sustainable development issues, including, for instance, the environment, industry, agriculture and tourism, and education, science and culture. The Programme is a good strategic document for the national level and a clear guide for the Ministry for the next seven years.

However, it does not include cost estimates or measures for financing. Consequently, its performance will depend on the development of projects that are able to attract funding from a variety of sources and on sector-specific plans. This includes the preparation of a second NEAP and, possibly, local environmental action plans (LEAPs).

National Programme for the Restoration and Expansion of Forests (2003)

Along with the National Programme on Environmentally Sustainable Socio-economic Development, presidential decree 1152, Approving the National Programmes on Ecology, also endorsed the National Programme for the

Restoration and Expansion of Forests. It lists activities in ten subsectors, along with indicative data on implementation, responsible institutions, financial sources and performance indicators. This Programme represents one of the few examples of a reasonably descriptive sector plan in Azerbaijan.

Partnership and Cooperation Agreement (1999)

Azerbaijan's Partnership and Cooperation Agreement with the European Communities has been in force since 1999. This Agreement represents another strategic document which effectively creates a set of international obligations on Azerbaijan, some of which are related to the environment.

In its preamble, the Agreement proclaims the will of its Parties to establish "close cooperation in the area of environment protection taking into account the interdependence existing between the Parties in this field". Article 56 establishes aims and means of cooperation on environment. The stated objectives are consistent with both the NEAP and the National Programme on Environmentally Sustainable Socio-economic Development. More importantly, its paragraph 3 mentions, inter alia, two strategic objectives, namely:
• Improving laws to European Community standards;
• Developing strategies, particularly with regard to global and climatic issues and to sustainable development.

The first of these two defines the harmonization of Azerbaijani environmental legislation with that of the European Union. If enforced properly, this clause will result in a far-reaching overhaul of the whole system of environment protection in Azerbaijan – a long-term objective. The second objective underlines the necessity to employ strategic planning as a tool to achieve the broad objectives of sectoral policy. Noticeably, the National Programme on Environmentally Sustainable Socio-economic Development states that developing regulations meeting European standards is one of the important conditions for meeting sustainable development goals.

National Caspian Action Plan

In May 1998, the five littoral States of the Caspian Sea adopted the Caspian Environment Programme. Pursuant to this Programme and its Strategic Action Programme, Azerbaijan is now in the process of developing a National Caspian Action Plan. (See chapter 8, on selected Caspian Sea issues.)

1.3   Legislative and Regulatory Framework

Since independence, and especially within the past two years, Azerbaijan has taken significant steps to improve the legal framework in the environmental sector and to strengthen its institutions. Nonetheless, many challenges remain to guarantee the constitutional right of citizens to live in a healthy environment.

Azerbaijan has adopted a structured procedural framework for law-making, stipulated in the Constitution and the Law on Normative-Legal Acts.

Theoretically, the public can take part in the law-making process; in practice, the extent of its participation depends on the leading institution (*Milli Mejlis*, or relevant government agency). According to the Law on Normative-Legal Acts, *Milli Mejlis* and appropriate bodies of the executive power compose plans of legislative acts that should be drafted and adopted. A body in charge of developing the draft law, as a rule, creates a commission consisting of its own employees, outside experts and scientists. Representatives of various stakeholders – NGOs, local authorities, scientists and State agencies – may be involved in the work at this stage. Law drafting may also be contracted to State agencies, academic and scientific institutions, NGOs or physical persons, including tendering for the best project. The leading agency may publish the draft law and/or initiate national discussion; however, there is no obligation to do so.

The Constitution

The Constitution was adopted on 12 November 1995 by referendum. It came into force on 27 November 1995 and was amended on 24 August 2002. According to the Constitution, Azerbaijan has a standard division of State power between the legislative, executive and judicial branches, represented respectively by *Milli Mejlis* (parliament), the President (head of State) and the courts.

The Constitution also addresses the decentralization of decision-making. The most important clauses refer to the status of Nakhchivan Autonomous Republic (arts. 134-138) and of local governments and municipalities (arts. 142-144). Nakhchivan is part of Azerbaijan, but enjoys a high degree of autonomy, having its own Constitution, elected parliament (*Ali Mejlis*) and separate government. Yet the Constitution of Azerbaijan, its laws, presidential decrees and resolutions of its Cabinet of Ministers are also binding on Nakhchivan.

The Constitution sets the following hierarchy for Azerbaijan's normative-legal acts:
- The Constitution;
- Acts adopted via referendum;
- Laws;
- Decrees;
- Resolutions of the Cabinet of Ministers;
- Normative acts of central executive bodies.
- International agreements and treaties ratified by Azerbaijan; if these contradict national normative-legal acts, acts adopted via referendum take precedence, but not the other laws.

The Constitution declares general principles, laying the foundation for the development of national environmental policy. Article 39 (Right to live in healthy environment) contains provisions regarding environmental quality, the right to information and environmental liability: "Every person shall have the right to live in a healthy environment. Everybody shall have the right to access information on the environmental situation and to get compensation for damage to health and property due to the violation of ecological rights." Article 78 (Environmental protection) stipulates that "the protection of the environment shall be the duty of every person."

According to the Constitution of Azerbaijan, local government in rural areas and towns, villages and settlements should be exercised by municipalities. Along with several other tasks, municipalities "approve and implement local ecological programmes", but they depend on "the necessary financial means for executing this authority. Realization of this authority is controlled by the legislative and executive power."

Environmental Laws

Azerbaijan has adopted many environmental laws and regulations covering different aspects of the sector as well as inter-sectoral relations. (see figure 1.1) In addition, the Law on Normative-Legal Acts stipulates in its article 8 that all acts in force prior to independence that were not subsequently cancelled and do not contradict the Constitution remain in force. However necessary for the transition period, this does make for a complicated mix of old and new regulations.

General laws

The *Law on Environmental Protection*, 8 June 1999 (latest amendment 30 March 2001), is the main piece of national environmental legislation. Its detailed and descriptive nature make it possible in some cases to apply it directly without sub-laws; this may be an advantage for a country in transition that cannot afford to draw up all the necessary normative documents in a short period of time.

The Law addresses the following issues:
- The rights and responsibilities of the State, the citizens, public associations and local authorities;
- The use of natural resources;
- Monitoring, standardization and certification;
- Economic regulation of environmental protection;
- Ecological requirements for economic activities;
- Education, scientific research, statistics and information;
- Ecological emergencies and ecological disaster zones;
- Control of environmental protection;
- Ecological auditing;
- Responsibility for the violation of environmental legislation;
- International cooperation.

Chapter VIII of the Law plays an especially important role in practice as it defines the principles and procedures for conducting ecological expertise. (See below.)

Figure 1.1: ADOPTION OF LAWS

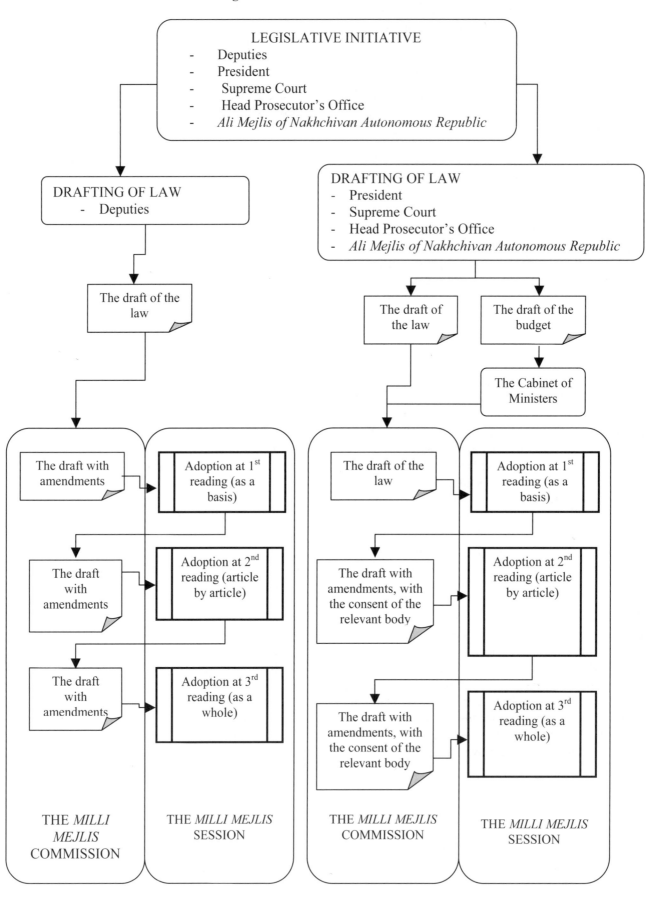

Graph after the VESCC. www.vescc.com

Ecosystem- and resource-specific laws

The *Law on Specially Protected Natural Areas and Objects*, 24 March 2000, determines the legal basis of the organization and preservation of protected natural areas and objects in Azerbaijan.

The *Law on Wildlife*, 4 June 1999, defines the animal world, property rights over fauna and legal relationships between actors. It also describes issues of State inventory and monitoring, and economic and punitive regulations.

The *Law on the Protection of Plants*, 3 November 1996, established the State Service for Plant Protection with responsibilities for the sanitary control and use of chemical, biological and other forms of plant protection.

The *Forestry Code*, 30 December 1997, creates the basis for the use, protection and restoration of forests. It establishes the types of forest ownership and property rights, the rules of management, and the responsibilities for the use and protection of forests. The Code also stipulates general administrative and criminal responsibility for the violation of forest regulations. (See chapter 9, biodiversity and forest management.)

The *Law on Mineral Resources*, 13 February 1998, regulates the exploration, rational use, safety and protection of underground resources and the Azerbaijani sector of the Caspian Sea. The Law lays down the principal property rights and responsibilities of users. It puts certain restrictions on the use of mineral resources, based on environmental protection considerations, public health and economic interests.

Provisional use of mineral resources may be for geological exploration up to 5 years, for production up to 25 years or for special use up to 30 years. The Law also sets general rules for the economic use of mineral resources and for liability for infringement of the Law.

Medium-specific laws

There is no legislation on soil protection, but the *Land Code* (1999), which is a comprehensive document, divides land into seven categories according to its use and legal status: agricultural land, residential area (settlements) land, land destined for industrial, transport, military, communication and other use, land of specially protected areas, land of the forest fund, land

of the water fund and land of the reserve fund. Only the State or municipalities may own land. Municipalities have the legal obligation to protect land in their possession, but in practice – owing to the lack of technical documentation and procedures – they are often not fully aware of what land they own or are unable to ensure its protection. (See also chapter 10, on land use, agriculture and desertification.)

The *Water Code*, 26 June 1997, regulates the use of water bodies, setting also property rights and covering issues of inventory and monitoring. State, municipalities and individuals may own water bodies depending on their importance. The Code regulates the use of water bodies for drinking and service water and for medical treatment, spas, recreation and sports, agricultural needs, industrial needs and hydro energy, transport, fishing and hunting, discharge of waste water, fire protection, and specially protected water bodies. It provides for issues of zoning, maximum allowable concentrations of harmful substances and basic rules of conduct for industry. (See also chapter 7, on water management.)

The *Law on Air Protection*, 2001, establishes the legal basis for the protection of air, thus implementing the constitutional right of the population to live in a healthy environment. It stipulates the rights and obligations of the authorities, legal and physical persons and NGOs in this respect, sets general requirements for air protection during economic activities, establishes rules for the State inventory of harmful emissions and their sources, introduces general categories of breaches of the Law that will trigger punitive measures. (See also chapter 5, on air management and transport.)

Health and safety laws

The *Law on Environmental Safety*, 8 June 1999, is one of two keystone laws of the country's environmental legislation (along with the Law on Environmental Protection). Its purpose is to establish a legal basis for the protection of life and health, society, its material and moral values, environment, including atmospheric air, space, water bodies, mineral resources, natural landscapes, plants and animals from natural and anthropogenic dangers.

The Law assigns the rights and responsibilities of the State, citizens and public associations in ecological safety, including information and liability. The Law also deals with the regulation of economic activity, territorial zoning and the alleviation of the consequences of environmental disasters.

This Law is primarily oriented to economic activity. In fact, its scope ("regulating relationships in the field of ecological safety during the implementation of activities" of legal and physical persons and authorities) fits within the broader scope of the Law on Environmental Protection ("regulating relationships between society and nature"). The latter explicitly cites "guaranteeing ecological safety" as first among its goals. Both laws were adopted on the same day but no attempt was made to integrate them. In general, to simplify the legal framework, it could be useful to combine the two into a single document, but there is no pressing need for this.

The *Law on Sanitary-Epidemiological Services,* authorized by presidential decree 371, 10 November 1992, aims to protect the population from the negative influence of the environment. It addresses the rights of citizens to live in a safe environment and to receive full and free information on sanitary-epidemic conditions, the environment and public health.

The Law sets out the rights and responsibilities of national (central) and local agencies, and prescribes basic sanitary requirements for certain subsectors of the economy. It also establishes four types of sanitary supervision: State sanitary supervision (national level), agency supervision (sector level), industry control (enterprise level), and public sanitary control.

The *Law on Public Health*, 26 June 1997, sets out basic principles of public health protection and the health care system. Special provisions are made for those living in a poor environment. The Law also assigns liability for harmful impact on public health, stipulating that damage to health that resulted from a polluted environment shall be compensated by the entity or person that caused the damage. (See also chapter 12, human health and environment.)

The *Law on the Radiation Safety of the Population*, 30 December 1997, outlines the legal basis for safe activities connected with radiation. To guarantee the safety of the population, the Law outlines measures that should be taken by authorities as well as by the users of radiation technology, including measures for public

information and education. The Law establishes State management and control of radiation safety and provides rights to the public to exert control on compliance by enterprises and organizations with rules and norms. The population and public associations also have the right to receive accurate information about the conditions and measures of radiation safety from the enterprises and organizations that use sources of ionizing radiation.

Waste and chemicals management laws

The *Law on Industrial and Municipal Waste*, 30 June 1998, describes State policy in environmental protection from industrial and household waste including harmful gases, waste water and radioactive waste. It defines the rights and responsibilities of the State and other entities, sets requirements for the design and construction of waste-treatment installations, and for the storage and transport of waste. There is also a general description of responses to infringements. (See also chapter 6, management of waste and contaminated sites.)

The *Law on Pesticides and Agrochemicals*, 1997, regulates the testing and registration of pesticides and agrochemicals, and defines the organization of agrochemical services. (See also chapter 10, land use, agriculture and desertification.)

Liability laws

The *Law on Mandatory Environmental Insurance* (2002) defines basic principles of insurance for activities that represent a risk to the environment. However, there seem to be many practical obstacles to its implementation. Recently, the Prime Minister's Order No. 38, dated 21 March 2003, established the list of the types of compulsory insurance, including also compulsory environmental insurance. The Prime Minister urged the appropriate governmental authorities to define the legislative basis and procedures for insurance.

The *Administrative Code* contains a chapter (9) on administrative infringements violating regulations of environmental protection, use of nature and environmental safety. Its articles 76-113 include a detailed list of violations and punishments. Punishments include fines measured in "minimum salaries." For the same violation, three different fines are applied – for physical persons, administrative persons (officials) and legal persons. The amount of the fine does not depend on the damage inflicted.

The *Criminal Code* (30 December 1999; in force since 1 September 2000) addresses "ecological crimes" in its chapter 28 (arts. 247-261). Punishments include fines, community service and imprisonment, or a combination of these. The punishment depends on the scale of the inflicted damage and can be applied to physical persons only.

Monitoring and information laws

The *Law on Access to Environmental Information*, 12 March 2002, establishes the classification of environmental information. If information is not explicitly classified "for restricted use", then it is open to the public. Procedures for the application of restrictions are described.

The *Law on Hydrometeorology*, 17 April 1998, establishes the basic concepts of State regulation and management of hydrometeorology as well as of the information policy on hydrometeorology and the natural environment. (See also chapter 3, environmental information and public participation.)

1.4 Other regulatory instruments

Regulatory instruments may provisionally be classified as following:
- Economic instruments;
- Monitoring and evaluation instruments;
- Impact assessment instruments;
- Environmental management instruments applied in organizations.

This chapter concentrates on the last two. (The first two are discussed in chapter 2, on economic instruments environmental expenditures and privitization, and chapter 3, on environmental information and public participation.)

Impact assessment instruments

Environmental impact assessment (EIA) (see figure 1.2) and strategic environmental assessment (SEA) fall in this category. In Azerbaijan, EIA is applied under the State ecological expertise (SEE) procedure; in contrast, SEA has never been applied, even though it has been introduced through SEE legislation.

State ecological expertise (SEE) and environmental impact assessment (EIA)

The Law · on Environmental Protection defines ecological expertise as "the identification of conformity of the environmental conditions with qualitative standards and ecological requirements in order to identify, prevent and forecast the possible negative impact of an economic activity on the environment and related consequences". State ecological expertise (SEE) applies to a very broad range of products and services, and even to their import (art. 52). The Law states that SEE is guided, inter alia, by international legal obligations.

The scope of SEE (art. 54) covers seven different applications. Of these, only three are fully consistent with the general concept of EIA. They are:
- Documentation relating to the development of new equipment, technologies, materials and substances, including those imported from abroad;
- Feasibility studies (calculations), construction projects (reconstruction, expansion, new technical equipment) and closing-down of structures and facilities, environmental impact assessment (EIA) documentation;
- The evolution of environmental conditions as a result of economic activity or emergencies.

Others would generally be considered more appropriate for strategic environmental assessment (SEA), for example:

- Draft State and local programmes relating to industrial development and siting at the national level and by economic sectors; and
- Regional environmental conditions and conditions of individual nature assets and complexes (ecosystem).

Two others, drafts of instructional and normative-technical documentation on environmental protection and environmental provisions of draft agreements (contracts), of relevance to natural resource use on the basis of a decision of the relevant body of the executive power, could be seen as subject either to EIA or SEA, depending on the context and content.

Neither this Law nor any other legal document gives any threshold values of activities that would require (or be exempt from) SEE. In theory, this means that all the applications mentioned above are subject to SEE. The Ministry of Ecology and Natural Resources is the responsible authority for SEE.

EIA, as a part of SEE, is, in fact, only required for development activities. However, the EIA legislation does not provide specific screening project categories. Consequently, all development proposals submitted to the relevant authorities for approval are subject to EIA. This means that it depends on the Ministry whether a project requires no EIA at all or a full EIA or anything in between.

Two normative documents on the conduct of EIA are in force in Azerbaijan: the 1990 Soviet Instruction on the Procedure of Estimation of EIA for Feasibility Studies and Projects, and Construction of Economic Objects and Complexes and the 1996 Azerbaijani Handbook for the EIA Process. (See figure 1.2.)

The regional departments of the Ministry of Ecology and Natural Resources receive applications and ensure that adequate information has been provided. Where EIA is required, documentation is sent to the head office of the Ministry for processing due to a lack of capacity in local offices. The Ministry must decide within one month on the extent of the EIA required, based on initial public enquiries within the locality of the proposed development and consultations with experts. For projects requiring a full EIA, the Ministry organizes and chairs a special scoping meeting of representatives of the applicant, invited experts and invited members of the public.

For each EIA, an environment review expert group is set up with skilled and experienced members (e.g. members of the Academy of Science, university staff or officials from other ministries). There are no firm requirements on group composition; the Ministry has a roster of experts, and composes each commission based on case-specific considerations. The expert group, with the participation of the public, makes recommendations to the Ministry, which then decides on whether to refuse the application or approve it with or without conditions.

Following a positive SEE, project proponents establish a monitoring programme to ensure that the conditions are not breached. The Ministry is responsible for verifying the accuracy and reliability of a proponent's monitoring results. It may issue a warning or halt an activity that

is breaching the conditions. If disagreements persist, the proponent has the option of taking the matter to the courts. Enforcement and compliance are the responsibility of the general inspection system.

In general, EIA works well in Azerbaijan. However, the lack of screening categories and fixed scoping requirements are a problem. While they allow greater flexibility, they also divert staff resources to smaller proposals, undermining the Ministry staff's ability to concentrate on major projects requiring a full EIA.

There are also other problems. Azerbaijani legislation requires project documents and EIA studies to be coordinated with other relevant institutions, but does not specify the form, purpose and time frame of this coordination. In evaluating alternatives, only technological alternatives need to be considered. Other kinds of alternatives, such as the "zero option" or no project, are not considered, nor are transboundary and global impacts.

Public participation is required for all stages of EIA and SEE. The general public and non-governmental organizations have the right to organize public ecological reviews for proposed projects, involving the Ministry in the process of information dissemination. So far, this right has not been used by any NGOs, possibly due to time and other resource constraints.

Public participation is required for all stages of EIA and SEE. The general public and non-governmental organizations have the right to organize public ecological reviews for proposed projects, involving the Ministry in the process of information dissemination. So far, this right has not been used by any NGOs, possibly due to time and other resource constraints.

Strategic environmental assessment (SEA)

As noted in the section on SEE and EIA, article 54 of the Law on Environmental Protection effectively calls for strategic environmental assessment (SEA) without mentioning it explicitly. SEA has been formally adopted in few countries in the region, and it is even more rarely mandatory. In this regard, Azerbaijani legislation seems quite progressive, but the reality is less optimistic. The SEA requirement of the Law on Environmental Protection is not supported by any sub-normative acts defining the procedures for its application or mechanisms for close cooperation between the Ministry of Ecology and Natural Resources and other State planning institutions. Not surprisingly, there have been no SEA applications.

1.5 Environmental Management Instruments

Environmental management instruments can assist in compliance promotion, and reduce direct costs. The most appropriate for economies in transition are, for instance, cleaner production, environmental labelling and environmental management systems, as their application can improve relations with authorities and decrease direct costs at the same time. Other instruments, such as life-cycle assessment, may be of interest in the longer term.

Azerbaijan's national development priorities are clearly oriented towards improving the investment climate and business regulations. Introducing environmental management instruments in the legal system would help to improve the investment climate. To date, none of the instruments mentioned above has been adopted in Azerbaijan.

1.6 Compliance and Enforcement Mechanisms

Permitting system: ecological passports

The system of permitting in Azerbaijan has its roots in the Soviet era; in fact, few – if any – changes have been introduced in the system for the past decade.

The statutes of the Ministry of Ecology and Natural Resources authorize it to issue ecological documents on the impact on the environment to potentially polluting enterprises. The documents include maximum allowable emissions, maximum allowable discharges, and an "ecological passport". This last item is specific to countries of the former Soviet Union.

At present in Azerbaijan, the ecological passport remains the main environmental document for enterprises. It contains a broad profile of an enterprise's environmental impacts, including resource consumption, waste management, recycling and effectiveness of pollution treatment. Each entity has to have one. Enterprises develop the draft passport themselves and then send to the Ministry of Ecology and Natural

Resources for approval. Approval is valid for three years, after which it must be reconsidered. Reconsideration is also required when there is any reconstruction or any other change requiring SEE.

Inspectors from the Ministry check compliance with the conditions set in the ecological passport. Enterprises pay both for "allowed" emissions and for excess emissions, using a three-tier system, including: payment for allowable units of pollution (normal rate); payment for each unit above the limit (increased rate), and a fine for breaches of the law plus compensation for the environmental damage done. All payments go to the State Environmental Protection Fund. Private enterprises and persons may also seek compensation from polluters in court.

The current design of the standard-setting system (including ecological passports) does little to encourage technological improvement. Only end-of-pipe solutions are encouraged and payment rates are low.

Compliance monitoring and enforcement

Enforcement can be effective only if the penalties outweigh the "benefits" of breaking the law, and this depends on three parameters: the rate of detection of violations (compliance monitoring); the rate of punished violations; and the penalty. Weakness in any of these three immediately leads to inadequate enforcement. In Azerbaijan, the first parameter depends almost solely on the work of the Ministry of Ecology and Natural Resources; the second, on the Ministry and others (police, courts); and the third, on courts and relevant legislation (Administrative Code, Criminal Code).

The Ministry of Ecology and Natural Resources is responsible both for the management of the environment and the State control of ecological safety. Inspection is therefore within its purview.

The organization of the inspection functions, however, raises questions regarding their efficiency and effectiveness. The State Control Inspectorate for Environment and Natural Resources is completely centralized. At the same time, the Ministry's Department of Environmental Policy and Environment Protection oversees 28 regional departments, which also have inspection functions. (see table 1.1)

Figure 1.2: EIA Process

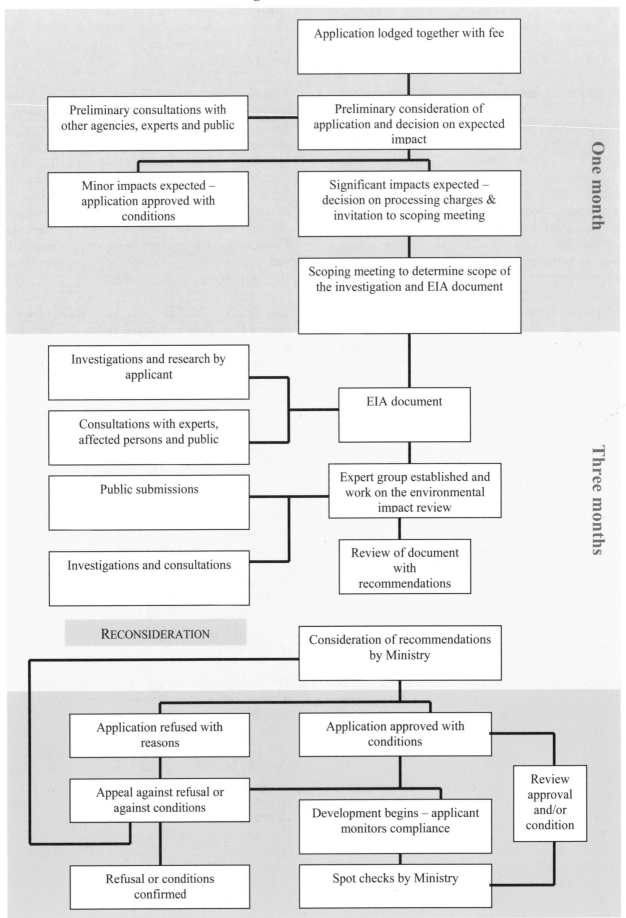

Table 1.1: Compliance monitoring and enforcement

Hierarchy of non-compliance responses	Authorized agency
(1) Informal responses, such as warning phone calls and letters with compliance recommendations	State Control Inspectorate, regional departments
(2) Formal warning letters and notices of violation	State Control Inspectorate, regional departments
(3) Financial penalties (fines), administrative or judicial, accruing as long as the violation persists	State Control Inspectorate, regional departments
(4) Suspension or cancellation of the permit(s), facility shut-down, forced corrective actions, or attachment of property, all of which may be required by administrative or judicial order	State Control Inspectorate, regional departments
(5) Criminal punishment, including imprisonment	Courts – case brought by the Ministry of Ecology and Natural Resources or other stakeholders

Source: Ministry of Ecology and Natural Resources, 2003.

The State Control Inspectorate is located in Baku and employs 110 inspectors in six divisions. It enjoys broad authority, including the right to impose administrative fines (according to the Administrative Code) and compensation for environmental damage (according to a decree of the Cabinet of Ministers of 1992, with subsequent amendments).

The State Control Inspectorate fields teams of inspectors according to a plan that, since September 2002, has to be approved by the Ministry of Economic Development. This measure was taken to protect businesses from "excessive" inspections. The Ministry of Ecology and Natural Resources sends the plan to the Ministry of Economic Development two or three times per quarter, well before the inspections take place. The Ministry of Economic Development can refuse some or even all the inspections, but, to date, all have been endorsed.

The regional departments of the Ministry of Ecology and Natural Resources have almost identical rights of inspection as the State Control Inspectorate, with some exceptions (e.g. in forests). Unlike the State Control Inspectorate, the regional departments also have other responsibilities as the regional representatives of the Ministry: they receive (but do not process) applications for SEE, interact with the local authorities on issues requiring the Ministry's approval or consultations, and take samples of environmental media and analyse them at their laboratories. Yet inspections constitute the bulk of their daily

work, for example, in Sumgayit, the regional department carried out 307 inspections in 2002.

The relationship between the inspection functions of the two is not clear, nor is the differentiation of their inspection responsibilities. The State Control Inspectorate has a right to inspect regional departments. Further, its employees are officially "inspectors", while those of the regional departments are "specialists".

In both cases, as is common in countries in transition, capacity is weak. Low salaries (equivalent to about US$ 45 per month), lack of equipment and lack of normative regulations are serious obstacles to enforcement. Regional departments tend to be the weaker of the two with respect to inspection.

The value of having two separate entities carrying out inspections is questionable for several reasons:
- There is less opportunity and fewer resources available to strengthen the capacity of the inspectorate;
- The management hierarchy for the inspections is different, although both report to the Ministry. Each drafts its own plans for inspections, creating the need for additional coordination to avoid overlap. This also means an unnecessary link in the enforcement chain;
- Response to the violations may require specialized actions – e.g. when bringing cases to court. This capacity and experience is easier to build within one centralized agency than at the local level;
- The overall burden on the regional departments is likely to increase as decentralization of power in Azerbaijan moves forward. In

particular, their role in SEE and permitting may grow. It would be logical to separate permitting and inspection operations institutionally before these changes take place.

1.7   Institutions responsible for the administration and implementation of environment-related objectives

The Ministry of Ecology and Natural Resources

The Ministry's statutes were adopted by presidential decree on 18 September 2001. Created on the basis of the former State Committee on Ecology, the Ministry took over the functions of several other State bodies: the Departments of Hydrometeorology, Geology, Forestry and Fishery. Therefore, the Ministry's employment structure looks asymmetric – out of 9,500 employees, only about 900 (incl. 500 at the local levels) work for environment divisions against 2000 for forestry.

The main responsibilities of the Ministry include:

- Implementation of State policy on natural resources research, use, restoration, protection and safety assurance, and biodiversity conservation;
- Implementation of State policy on the use of bioresources of both internal water bodies and the Caspian Sea (although not of the water resources themselves);
- Implementation of State policy on geological exploration, mineral resources protection and use;
- State administration in the field of the environment;
- Organization of hydrometeorological services;
- Implementation of State control over ecological safety compliance;

- Within its competence, implementation of relevant international agreements and coordination of other bodies in this field.

Unfortunately, the Ministry was unable to function properly for many months after its establishment owing to the usual burden of restructuring but also owing to the late availability of funding in 2002. Its annual budget is about US$ 8 million.

Its present structure (see figure 1.3) includes several departments at the central office, and a large number of subordinated facilities (e.g. fish hatcheries, seedling productions, hydrometeorological units). Coordination among so many different units requires well-developed procedures and sub-normative documents that must be upgraded over time.

The structure of the Ministry continues to bear signs of the constituent parts that were brought together to create it. For example, the geology, hydrometeorology, forestry and fishery sub-sectors have separate departments, as does the environment in general. Consequently, some tasks are divided among several departments; for instance, there are three different departments with monitoring responsibilities. The Ministry, therefore, appears still to be in a transitional stage from a complex combination of pre-existing organizations to an integrated and manageable institution. It is expected that further adjustments to the Ministry's structure will be unavoidable in the future.

The Ministry of Ecology and Natural Resources also has a network of 28 regional departments for environment and natural resources subordinated to its Department of Environmental Policy and Environment Protection. (See section 1.6 above.)

Figure 1.3: Organigram of the Ministry of Ecology and Natural Resources, Azerbaijan

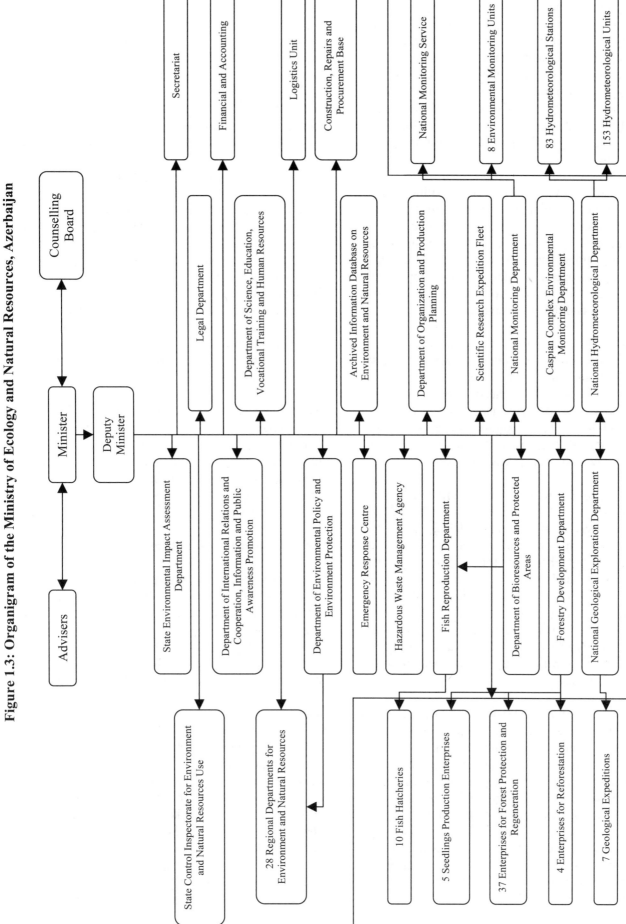

Decentralization

Azerbaijan is divided into administrative districts (rural *rayons* and cities). According to the Constitution, local governance in Azerbaijan is exercised both through local bodies of State administration (executive powers) and through municipal governments. The former were established through the Constitutional Provision on Local Executive Power. Municipal government responsibilities are prescribed by various laws.

About 200 city (town) municipalities and 2,500 rural (village) municipalities were established through elections in December 1999. Municipalities are entitled to establish local taxes and duties, adopt a local budget, manage municipal property, adopt and implement programmes for social protection, economic development and the environment.

In fact, the executive powers, which are subject to centralized authority, wield almost all power on the local level. According to the Constitution, their heads are appointed and removed by the President, who determines their authority. The regional departments of the Ministry of Ecology and Natural Resources work closely with the executive powers.

As the State Programme on Poverty Reduction and Economic Development for 2003-2005 indicates, there is still no clear definition of the role and functions of municipalities, no clear distinction between the authority of the municipalities and that of the local executive bodies and no clear definition of their relationship with the central government agencies. Municipalities still do not have full use of the property that is allocated to them, nor the funding which they are entitled to from the central budget.

Local executive powers existed also in the former Soviet Union, where they represented local branches of central government with the clear vertical subordination and minimum independence in decision-making. In Azerbaijan, they retain these characteristics. The inability of municipalities and limited initiative of local executive powers creates little prospect for local environmental initiatives like LEAPs or specific community-level programmes, and effectively leaves the Ministry's regional departments solely responsible for local environmental problems.

Coordination and integration

While the Ministry of Ecology and Natural Resources has primary responsibility for environmental management, other ministries and committees also have important functions that directly relate to those of the Ministry. These include, in particular, the Ministry of Health, the Ministry of Agriculture, the Ministry of Economic Development, the Ministry of Fuel and Energy, the State Committee of Amelioration and Water Management and several other State committees (standardization, industrial safety).

The Cabinet of Ministers is officially responsible for coordinating government agencies. But, at the working level, there are no mechanisms for coordination and integration. This is especially a problem for the development of the implementation of key programmes. There is a danger, for example, that the Ministry of Ecology and Natural Resources is expected to implement the National Programme on Environmentally Sustainable Socio-economic Development and the NEAP on its own.

Regarding coordination between the Ministry of Ecology and Natural Resources and interested non-governmental organizations, the Minister took the initiative in 2002 to begin a structured dialogue with other stakeholders on a regular basis. (See chapter 3, environmental information and public participation.)

1.8   Conclusions and Recommendations

After 10 years of independence, Azerbaijan continues to describe itself as "a nation in transition to democracy." The basic political and legal parameters for the institutionalization of democracy have been established and are being refined and enacted. This process involves dismantling institutions, revising laws and defining new ones to bolster an open, market-oriented society.

While the environment is protected by law and pollution is controlled by regulations, in fact, concern for the environment has been secondary to economic development. It is therefore important that environmental legislation and management should be given a higher priority to meet the future needs of Azerbaijan.

There are clear signs that priorities are changing and more attention is given to the environment:

radical institutional change has brought to life the Ministry of Ecology and Natural Resources, which – despite serious obstacles – has been able to take a lead and push the environment higher on the list of national priorities for action. It is very important for the Ministry as well as for Azerbaijan's environment not to lose that momentum. Decisions recently adopted in Kiev, at the fifth Ministerial Conference "Environment for Europe", provide a good international climate for reforms and reliable guidance for improvements. Of course, Azerbaijan's ambitions should be kept within the limits of its possibilities; yet these possibilities have to be reassessed and they will, most likely, grow over time.

When implementing the recommendations given below, the Ministry of Ecology and Natural Resources can use the potential of international organizations that have prepared many sound studies and background analyses applicable to Azerbaijan as well. Also the potential of cooperation under the "Environment for Europe" and other processes is not fully used.

Azerbaijan has been active in formulating policies for the environment, for sustainable development, and for poverty and economic development. Within the first two years of its establishment, the Ministry of Ecology and Natural Resources had prepared four national programmes, two of which were approved by President's Decree in February 2003, and discussions have taken place with other ministries on their implementation. The other two programmes have been submitted to the Cabinet of Ministers. However, the relationship among these programmes, and their relative priority, is not always clear, and there is not yet a plan for their monitoring, review and revision. In addition, the Ministry of Ecology and Natural Resources is the main body that initiates environment related activities, which also bears responsibility for their implementation. It is, however, impossible to implement them without good coordination among all government institutions, integration of environment into other sectoral policies and plans and provision of adequate funding. The environmental planning process would benefit from a more consolidated and rationalized framework.

Recommendation 1.1:
The Ministry of Ecology and Natural Resources, in consultation with other relevant institutions, should prepare and submit through Parliament or to the President, an implementation programme for a continuing process of policy-making for environment and sustainable development. This programme should provide an overall framework for policy; establish a schedule for monitoring, reviewing and revising policies; and indicate the relationship and hierarchy among policies. The programme should be of a multi-sector nature, and not be limited to only the obligations of the Ministry of Ecology and Natural Resources. To the extent possible, it should also specify sources of financing for implementation.

To achieve the declared goals of national environmental policy, comprehensive environmental legislation is essential. Azerbaijan has progressed considerably in this field. Still, its legislation represents a mixture of old and new approaches, which is typical for many countries with economies in transition, but is neither efficient nor effective. In addition, a number of lower-level normative acts (rules, standards, procedures, regulations) that are vital for the implementation of general provisions of laws are lacking. The Ministry of Ecology and Natural Resources has begun the process of analysing the legislative gaps and has indicated its commitment to this process.

Recommendation 1.2:
The Ministry of Ecology and Natural Resources should continue and finalize its "gap analysis" of Azerbaijan's environmental legislation, with particular reference to the Partnership and Cooperation Agreement with the European Union and other internationally adopted principles. Conclusions of this analysis would provide the basis for the development of a Plan for Legislative Work in the Environmental Sector, together with the Milli Mejlis Commission on Environment and other stakeholders, especially national non-governmental organizations. The Plan should avoid being overambitious, and should take a step-by-step approach, sufficiently supported by growing human and financial resources throughout its implementation.

While policy planning and legislation are important, overall success can only be measured through implementation. The current system of State ecological expertise is described in the 1999 Law on Environmental Protection, and it applies to a very broad range of products and services, activities and policies. In this respect, it combines both environmental impact assessment and strategic environmental assessment in a single package, with no clear differentiation between them. It is important to update the system and make it consistent with standard international practice.

Recommendation 1.3:
The Ministry of Ecology and Natural Resources should undertake the following:
(a) Redesign the system of Ecological Expertise with environmental impact assessment legislation based on international experience and practices, with clear guidelines regarding screening and scoping procedures; initial steps towards decentralized decision-making in this area should be planned for the mid-term;
(b) Develop separate legislation for Strategic Environmental Assessment (SEA), which applies to a higher stage of national planning and requires a higher degree of coordination.

The Azerbaijani Government (including the Ministry of Ecology and Natural Resources) has a strong vertical administration that is well positioned for centralized implementation and enforcement. In the case of environmental legislation, compliance and enforcement responsibilities are mostly concentrated within the Ministry of Ecology and Natural Resources; its enforcement structures need to be better consolidated and empowered. This requires the development of new legal documents and procedures as well as adequate financing and human capacity. At the same time, the responsibilities of the central office and the regional divisions should be clearly delineated – particularly in the area of inspection.

Recommendation 1.4:
The Ministry of Ecology and Natural Resources should restructure the State Control Inspectorate

for Environment and Natural Resources (SCI), in an effort to:
(a) Consolidate central and regional inspections into a single system, with clear rules of procedure and differentiation of responsibilities. This would include placing the regional inspection functions under the State Control Inspectorate and removing them from the Ministry's Department of Environmental Policy and Environment Protection. The restructuring process should also evaluate the relationship between regional inspectors for environment and those for health; (see also Recommendation 12.1)
(b) Provide greater autonomy to the State Control Inspectorate and sufficient resources for it effectively to carry out its work; and
(c) Strengthen the capacity of the State Control Inspectorate for Environment through intensive training of inspectors and through the implementation of a national standardized and mandatory recruitment exam for all inspectors.

Recommendation 1.5:
The Ministry of Ecology and Natural Resources should assess the entire national framework for compliance and enforcement, with the aim of developing and implementing a well-articulated enforcement strategy, which should, inter alia:
(a) Identify the weaknesses in the present system of compliance and enforcement (e.g. absence of procedural documents, overlapping of responsibilities of various agencies, low level of financing and motivation, outdated standard- and payment-setting approaches, inadequate court proceedings) and prepare a list of legislative and institutional measures to address these problems. This list should form the nucleus of an action plan;

Give special attention to the use of compliance promotion measures, (e.g., cleaner technology centres, voluntary environmental audits, environmental management systems and eco-labelling) in parallel with compliance monitoring and enforcement, and to setting firm and transparent procedures for this.

Chapter 2

ECONOMIC INSTRUMENTS, ENVIRONMENTAL EXPENDITURES AND PRIVATIZATION

2.1 Introduction

Some economic instruments for the use of natural resources (e.g. hunting licences) and financial sanctions for violating environmental legislation on the protection of animal and plant species, forestry and fisheries were introduced in Azerbaijan as early as the 1980s. Revenues accrued in the Reserve Fund of Nature Protection, which was established in 1989 and governed by the former State Committee on Ecology and Control of Natural Resource Use under the office of the President.

A comprehensive system of economic instruments was designed in 1992 mainly to protect the environment and ensure the rational use of natural resources. The Law on Nature Protection and Environmental Management of 25 February 1992 laid down the legal basis for the implementation of the polluter/user-pays principle. The Cabinet of Ministers established procedures and rates for charges and fines on air emissions, waste-water discharges and solid waste disposal, as well as fees for the abstraction of surface and groundwater (Resolution 122 of 3 March 1992).

To counterbalance the eroding impact of high inflation, the Cabinet of Ministers increased the rates of most environmental charges and fees tenfold in 1993. Payment rates for the illegal use of mineral resources were increased in 1998. Since then, the rates have not been adjusted to inflation, which admittedly is now moderate.

Azerbaijan has also introduced taxes and excise duties, which capture some natural resource rents. Taxes and charges on natural resource extraction and energy taxes, particularly on oil, play an important role as a source of revenue for the State budget. Azerbaijan attempts to reinvest at least some portion of oil tax revenue in social and environmental infrastructure and diversification of the economy to make it less dependent on low-value-added natural resource exports.

Through price regulation Azerbaijan seeks to counterbalance the monopoly position of suppliers of goods and services that are crucial for economic and social policies. However, price control over petrol, electricity, heating and water supply restrains resource savings and associated environmental benefits.

The ongoing privatization is expected to improve resource efficiency and environmental accountability. Some aspects of environmental oversight are becoming more complicated, however, and the environmental authorities have to be actively involved in the privatization process to monitor such issues as liability for past environmental damage, environmental legal requirements of privatized enterprises and environmental commitments in bidders' investment plans.

2.2 Economic instruments

Environmental charges, fees and fines

There are two types of emission charges, namely charges for emissions of pollutants within the established limits and charges for emissions exceeding those limits. The latter are five times higher than the former. Fines are imposed on polluters for accidental emissions and environmental pollution caused in the absence of established emission permits. Fines are ten times higher than the respective charge rate for emissions within established limits. All charge and fine rates depend, furthermore, on the environmental conditions in a given area and its ecological significance.

If an enterprise causes damage to the environment, it should pay compensation. The guidelines for assessing environmental damage used by the environmental authorities when preparing compensation claims were developed in the former Soviet Union in the 1980s.

Legal persons guilty of non-payment of mandatory charges and fines may be taken to court. However, Resolution 122 allows local authorities, in agreement with tax inspectors and the relevant bodies of the Ministry of Ecology and Natural Resources, to consider the financial situation of enterprises and exempt them from charges for emissions above established limits or reduce their overall payments of pollution charges. Regional environmental committees may also allow enterprises that take agreed environmental protection measures to pay charges for excessive emissions at the rate of charges for emissions within established limits.

Air Pollution

The system of air emission charges is very similar to the system used in most other countries of Eastern Europe, the Caucasus and Central Asia. Charges, which are levied on 88 different pollutants, vary according to the degree of hazard of the pollutants. Charge rates range from virtually zero for several pollutants to 10.1 million manats per ton for the most toxic. The applicable charges vary among regions to reflect differences in environmental conditions. The base rate is multiplied by a regional coefficient between 1 and 5. The highest values are in Baku and in the Sumgayit area – both are hot spots. Basic rates of selected air emission charges are presented in table 2.1.

Table 2.1: Charge rates for the emission of selected air pollutants

(manats/ton)

Air pollutant	Rate
Nitrogen dioxide	329
Sulphur dioxide	132
Benzene	10,100,000
Hydrocarbons	20
Carbon monoxide	8
Ammonia	37
Benzene	105
Isoprene	59
Nitric acid	226
Arsenic	12,648
Carbonate of sodium	784
Nickel	43,800
Mercury	214,664
Soot	332
Lead	172,000
Hydrogen sulphide	329
Chlorine	715
Phenol	136

Source: Government Resolutions No. 122 of 3 March 1992 and No. 216 of 1 May 1993.

Another air emission charge is applied to mobile sources. The air emission charge on motor fuel is levied per ton of air emissions from fuel combustion. Based on standard emission factors, charge rates are calculated for different types of fuel. As emissions from diesel fuel, leaded and unleaded petrol differ, this creates a charge differentiation in favour of unleaded petrol. The charge rates for leaded petrol and diesel fuel are 61.7 and 50.1 manats, respectively, per ton of polluting substances from combusted fuel, whereas for unleaded petrol the charge is 37.4 manats/ton. This charge is so low however, that in practice there is no incentive to use unleaded petrol. Nevertheless, leaded petrol is no longer produced in Azerbaijan and its import is close to zero.

Water pollution and abstraction

The purpose of the waste-water charge is to motivate water users to comply with the sewage discharge limits and reduce their discharges of polluting substances into natural water bodies, either directly or through municipal sewerage systems. Waste-water charge rates are different for the Kura river basin and for basins of other rivers and the Caspian Sea. In all cases, however, the charge level depends on the amount of clean water required to dilute the pollution up to the established water quality standards. The base rate per m^3 of water required to dilute the waste water varies between 16 kopeks for the Kura river basin and 12 kopeks for basins of other rivers and the Caspian Sea.

Fees for surface water abstraction differ according to five water quality categories established for the Kura river catchment area and for the catchment areas of other rivers, respectively. Water quality categories are established for individual water bodies or parts of these according to the State Classification. Fee rates are presented in table 2.2.

The fee for the abstraction of groundwater in the Nakhchivan autonomous republic (26.4 kopeks/m^3) is approximately one third of that applied in the rest of Azerbaijan (71.5 kopeks/m^3).

Table 2.2: Fees for surface water abstraction

(kopeks/m³)

Water quality	Rate
Kura River basin	
I	34.3
II	33.5
III	31.7
IV	29.3
V	26.5
VI	21.7
Other river basins	
I	26.3
II	25.7
III	24.4
IV	22.6
V	20.5
VI	16.9

Source: Government Resolutions No. 122 of 3 March 1992 and No. 216 of 1 May 1993.

Waste disposal

A charge is levied on solid waste and the rate is differentiated according to the type of waste and its degree of toxicity. There are five classes of waste, including one class of non-toxic waste, for which the charge is 50 manats/ton, whereas the others range from 250 (slightly hazardous – class IV) up to 1,500 manats/ton for the most toxic class. The regional coefficients also apply to waste. Enterprises may be exempt from these payments if they discard their waste in specially equipped landfills that prevent emissions into air and water.

Mineral extraction

Fines on the non-licensed extraction of certain minerals depend on the mineral. The fines are defined for eight types of minerals, varying between 16,500 manats/m³ (rocks of various composition) and 275,000 manats/m³ (building sand). The fines are three times higher if the extraction takes place in protected areas and recreational zones.

Forestry

Logging charges are divided into stumpage fees and charges for firewood and industrial wood. The former is a fixed charge, and the latter is set in accordance with the wholesale price that is established by the Forestry Development Department under the Ministry of Ecology and Natural Resources. The aim of the charges is to stimulate the rational use of forest resources

and to generate revenue for the forest management authorities.

The current stumpage fees are defined according to four factors. First, a distinction is made for the timber species in question. Eleven groups of timber species are defined. The highest rate applies, for instance, to walnut, chestnut and plane trees. Second, rates for this group vary according to three quality categories and range from 165,322 to 46,298 manats/m³ of dense timber. Third, the rates are differentiated according to the required hauling distance (0-10 km, 10.1-25 km and 25.1-40 km). Fourth, the rates vary according to the size of the timber in question (small, average and large).

Payments are collected from timber and firewood sales. The revenue covers up to 40% of the yearly reforestation costs.

Biodiversity and nature protection

All legal and natural persons engaged in non-commercial hunting are liable to pay a hunting charge. The charge depends on the type of game and there is a quota on the number of animals that may be killed during one season. For instance, the charge for brown bear is 165,000 manats per animal, and for mountain goat it is 110,000 manats. For most other game, the rate is 10,000 manats per animal killed.

Legal actions are brought against those who hunt and catch wild animals and fish (in excess of the fixed limits or without licences), and fines are imposed. The rate depends on the species in question and is defined per unit of game hunted. It varies between 11,000 manats for quail and sandpiper, and 825,000 manats for mountain goat. For fish, the rates are defined per fish caught. Rates vary between 5,500 manats and 330,000 manats (for great sturgeon). The current fines do not provide the desired disincentives and, pursuant to the recent poaching of a leopard (a flagship threatened species in Azerbaijan), the Ministry of Ecology and Natural Resources has proposed to the Cabinet of Ministers to drastically increase the fines for a number of protected species.

Actual collection of revenues

Total revenues collected from environmental charges, fees, fines and compensation for damage are presented in table 2.3. A sharp decrease in revenue in 2001 is explained by the weakening of the environmental authorities, including the enforcement administrations in the country caused by the dissolution of the State Committee on Ecology and Control of Natural Resource Use that year. It took some time before its successor, the Ministry of Ecology and Natural Resources, became fully operational. The increased collection rate for 2002 demonstrates, inter alia, the strengthening of law enforcement in Azerbaijan rather than an increase in pollution.

In all 333,440,000 manats (some US$ 59,000) were collected over the first five months of 2003. Figure 2.1 shows that the main revenue sources were payments for air and water pollution This breakdown generally confirms the trend of recent years.

Charges for authorized emissions constitute the bulk (46%) of total revenues from environmental instruments, followed by compensation payments for environmental damage (23%), fines (21%) and payments for the violation of the status of nature reserves (10%). The actual collection rate remains low, however. For emission charges it is only 70%, while for fines it is 37% and for compensation payments only 13%.

Other economic instruments related to the environment

Taxation policy

A Tax Code entered into force in Azerbaijan on 1 January 2001. It lowered the overall taxation burden and simplified the system of calculation and collection. As a result, the tax collection level, particularly from businesses, has drastically improved. SOCAR, the State oil company, is the country's major taxpayer. In 2002, it paid 1,660 billion manats in taxes. That represented some 36% of the total levies to the budget of Azerbaijan. Oil extraction taxes and excise taxes on the production of oil and oil products represented the main taxes paid by SOCAR. Foreign oil companies are exempt, however, from these taxes.

New tax rates on mineral extraction or royalties have been in effect since 1996. Their main purpose is to raise revenue for the State budget. The tax is imposed on the wholesale price of the product. Tax rates are 26% for crude oil, 20% for natural gas and upto 10% for ore and non-metallic minerals. The

collected tax revenue amounted to some 250 billion manats in 2002 (about 7.4% of total tax revenue).

The excise taxes on petrol are rather low. They vary according to petrol quality: 144% for AI95-grade petrol, 135.7% for AI92-grade petrol and 144.9% for A72-76-grade petrol. The excise tax on lubricated oils varies from 70 to 80% depending on the oil type. No excise tax is levied on diesel fuel as the latter is mostly for export.

Producers pay the excise tax, which accrues to the State budget. The revenue from the excise tax on oil products constituted 98% of all excise taxes collected in Azerbaijan in 2002. The share of excise taxes in the overall tax revenue amounted to 13% in the same year.

All energy products are subject to the general value-added tax (VAT) at a rate of 18%. A collection rate of 100% was reported for VAT in 2002.

There are a number of duties and taxes that are levied on vehicles in Azerbaijan. Their main purpose is to collect revenue for the State budget.

Import duties on motor vehicles have been in force since 1995. The import duty is to be paid by all natural persons that import motor vehicles for private use. The duty is based on the engine capacity, and it is differentiated according to the manufacturing year. If the vehicle is less than one year old when it enters the country, the rate is US$ $0.4/cm^3$. If the vehicle is older, the rate is US$ $0.7/cm^3$. Payments must be made in manats although rates are set in dollars. The duty rate does not discourage the import of old (and, hence, environmentally less friendly) cars. The total annual payments amounted over the past few years to 40-50 billion manats. The collection rate is high (99.4%).

Table 2.3: Revenues of environmental economic instruments

(in thousands of manats)

Year	Amount
1997	462,113
1998	711,363
1999	813,567
2000	826,294
2001	391,794
2002	1,103,900

Source: Ministry of Ecology and Natural Resources, 2003.

Figure 2.1: Share of individual revenue sources, 2003

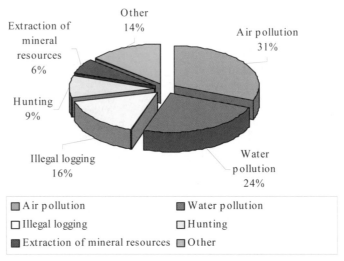

Source: Ministry of Ecology and Natural Resources, 2003.

Owners of vehicles and wheeled machinery have to pay an annual property tax on their vehicles and a service charge for preventive maintenance. The rates vary according to the engine volume. Neither the tax nor the service charge provides incentives to use low-consumption and low-emission cars.

The transit tax applies to the entry and transit of foreign-owned vehicles. For freight traffic, the transit tax is calculated based on the freight weight and hazardousness. The average tax is equivalent to US$ 20-30. Passenger traffic rates are US$ 15 for private cars and US$ 15-25 for buses.

The land tax was introduced in 1993. The current legislation governing the use of the tax entered into force in 1999. The tax applies to all local and foreign legal and natural persons who rent or own land in Azerbaijan irrespective of what they do with it. This tax is differentiated according to the area, geographical position and quality of the land. For rural areas the tax is defined per hectare of land, and for towns and populated areas per square metre. The rates are very low. Certain types of land are exempt, e.g. nature parks and land for common public use.

In 2002, the Ministry of Taxes collected some 44 billion manats in land taxes. The collection rate was about 80%. Revenue accrues to the State budget and, although it should primarily be used to finance land amelioration measures, the revenue is often used for other purposes. The tax does not serve any environmental function, however.

Price regulation and tariff setting by public authorities

Besides imposing taxes on oil and oil products, including petrol, the Government sets price limits on these products. This direct intervention keeps energy prices relatively low. The current retail price of petrol, for instance, is around 1,500 manats (US$ 0.3) per litre at present.

The Tariff Council of the Ministry of Economic Development regulates electricity prices. Tariffs are different for the three groups of established users: 96 manats per kWh for households, 130 manats for industry and State organizations, and 250 manats for commerce, trade and services. In spite of the cross-subsidization in favour of households, 60% of them do not pay their electricity bills. Electricity tariffs are linked to the price of heavy oil. The current tariffs have stayed unchanged since November 2000, although the price of heating oil doubled in February 2003. The next revision of electricity tariffs is expected to take place only in 2004.

The executive powers and municipalities regulate heating prices. In Baku, households pay 250 manats/m^2 per month for centrally supplied heat. In addition, the supply of hot water from one of the two State-owned power stations costs 700 manats per person per month. Tariffs for other users are calculated on a different basis. For the public sector, the monthly rate is 600 manats/m^3 of heated space, for the private sector, it is 1,100 manats. The monthly tariff for hot water is 96,000 manats per gigacalorie for both sectors. The current tariffs for heating were set in 1999. The money collected does not cover the full costs. Heating is subsidized for all users, albeit to a different extent.

Energy pricing in Azerbaijan does not provide incentives to energy producers to save costs in the production and supply of energy. Neither does it provide incentives to users to save energy. As a consequence, the economic growth that has recently started in Azerbaijan may lead to an increase in energy production and consumption and in their adverse environmental impact.

Nor does water pricing promote water conservation. The Tariff Council sets water prices. Table 2.4 presents the tariffs for water supply set for Baku-Absheron and several adjacent regions. Water supply is cross-subsidized in favour of households. For the population, the monthly water price is very low: the equivalent of US$ 0.5 per person. Nevertheless, owing to the low living standards, only 60% of households pay their bills. Money collection from other water users is similarly poor. In total, only 50% (some 130 billion manats) are collected annually from all water users.

Water cost recovery is even lower at other water-supply companies. The executive powers and municipalities heavily subsidize their operations and investment costs.

The executive powers and municipalities also set tariffs for municipal waste collection and disposal from households, enterprises and organizations. Tariffs are based on the cost incurred for the services provided. Tariffs for the City of Baku, major towns and other populated areas vary between 750 and 1200 manats/m³ of solid waste. The revenues seem to be sufficient to cover cost. However, the extent to which the collected revenue can finance the safe disposal of municipal solid waste is difficult to establish.

Table 2.4: Water tariffs for Baku-Absheron and several adjacent regions

(manats/m³)

User	Price
Drinking water for:	
Households	185
State institutions	900
Commerce and services	5,300
Beverage producers	42,000
Other producers	2,500
Technological water for production purposes and for big industrial consumers	900
Water for irrigation	4,000-14,000*

Source: Ministry of Economic Development of Azerbaijan, 2003.
Note: *Depending on water sources and land quality.

2.3 Environmental expenditures

State budget

The share of the reported environmental expenditure in Azerbaijan is 0.4% of gross domestic product (GDP). It is both people's low income and their unwillingness to pay that seem to be holding back higher environmental expenditures. According to the World Bank's definition (World Bank Global Development Finance), Azerbaijan belongs to the category of low-income countries whose annual gross national product (GNP) per capita is equivalent to $760 or less.

Domestic environmental expenditures in absolute terms in the 1997-2002 period are presented in table 2.5 and figure 2.2. The bulk of expenditure is allocated to short-term needs, mostly in water supply and waste-water treatment.

Table 2.5: Trends in total environmental expenditure, 1997-2002

(million manats)

	1997	1998	1999	2000	2001	2002
Expenditure						
Total	**92,920**	**99,087**	**86,272**	**91,294**	**104,161**	**84,690**
of which:						
Current expenditure	57,828	50,712	54,719	59,270	51,208	58,243
Maintenance of reserves, wild animals and their reproduction	728	1,308	1,063	1,358	1,027	1,365
Forestry	3,325	3,808	5,197	6,409	5,665	7,263
Investments	11,147	19,153	6,384	8,614	6,356	12,025
Repair and maintenance	19,892	24,106	18,908	15,645	39,260	5,794

Source: Data for 1997-2001. State Statistical Committee. Environment 2002. Data for 2002. Direct communication with Ministry.

Figure 2.2: Trends in total environmental expenditure, 1997-2002

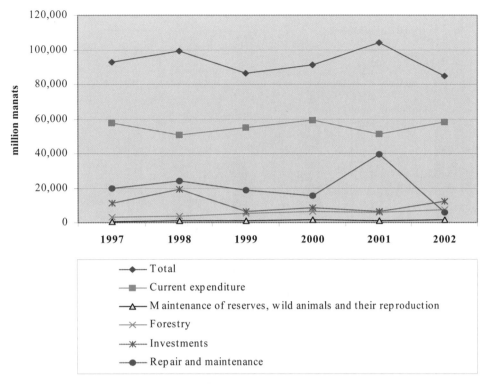

Source: Data for 1997-2001. State Statistical Committee. Environment 2002.
Data for 2002. Direct communication with Ministry.

Environmental investments in the public sector contribute 0.1% to total investments in the public sector of the economy. Table 2.6 and figure 2.3 present the share of environmental investments in total investments and the breakdown of environmental investments from 1997 to 2002.

The underlying reason for this persistently low level of environmental capital investment is the very weak demand from public authorities for improving environmental conditions.

The budget of the Ministry of Ecology and Natural Resources is itself insignificant. In 2002, it amounted to some 36.7 billion manats and in 2003 it was slightly increased to 39.9 billion manats. Nearly one third of the budget is earmarked for geological exploration, however. The budget of the Ministry is supplemented by revenues earmarked in the State budget for the State Environmental Protection Fund.

State Environmental Protection Fund

The 1999 Law on Environmental Protection established the principles for the operation of a State environmental fund. On 22 February 2001,

the Cabinet of Ministers adopted Resolution 41 by which it transformed the Reserve Fund of Nature Protection into the State Environmental Protection Fund. It established the objectives and the method of operation of the Fund and consolidated the Fund's revenues into the State budget.

The State Environmental Protection Fund is intended to finance environmental protection, including the creation and management of protected areas, to mitigate environmental damage, to support environmental projects, monitoring, education, research and development, and international cooperation, and to provide financial incentives for environmental improvements.

Table 2.6: Share of environmental investments in total investments in the public sector of the economy, 1997-2002

					(%)
1997	1998	1999	2000	2001	2002
0.2	0.3	0.1	0.2	0.1	0.1

Source: Data for 1997-2001. State Statistical Committee of Azerbaijan Republic. Environment 2002. Data for 2002. Direct communication with Ministry.

Figure 2.3: Environmental investments, 1997-2002

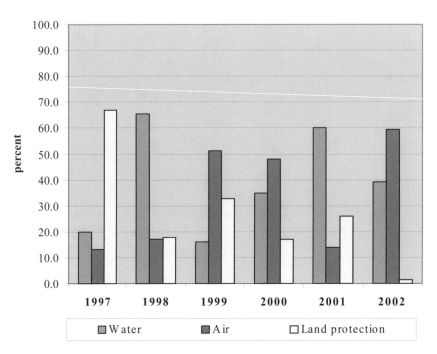

Source: Data for 1997-2001. State Statistical Committee. Environment 2002.
Data for 2002. Direct communication with Ministry.

Its revenues should come, primarily, from charges and fines for environmental pollution, fees for the use of natural resources, fees for hunting licences and licences for the export of medicinal plants (discontinued in 2003), fines for the violation of the protection status of nature reserves, and revenues from selling confiscated game and fish stocks and other illegally obtained natural products. The revenues should be used by the Ministry of Ecology and Natural Resources for purposes that are agreed annually with the Ministry of Finances and coordinated with the Ministry of Economic Development.

Revenues that are collected annually in the State Environmental Protection Fund are directed to environmental actions by the Ministry of Ecology and Natural Resources. In 2002, for instance, of the total 1,103.9 million manats received by the State Environmental Protection Fund, 974.15 million manats were transferred to the Ministry. In 1996-2000, the share of the Reserve Fund of Nature Protection in total public environmental expenditure accounted for only 0.5-1%. For the past two years, revenues to the State Environmental Protection Fund have increased by a factor of 1.5.

Foreign sources

The share of foreign money in total environmental expenditures (public sector only) over 1996-2001 accounted for some 39%. Over this period, Azerbaijan received annually 2,500 manats worth of environment-related assistance on average per inhabitant. The assistance came from a few multilateral and bilateral donors (see chapter 4, on international cooperation). Table 2.7 presents data on the volume of this assistance over the said period.

To raise additional donor funds, project proposals could be developed aimed at solving priority environmental problems.

2.4   Privatization

Privatization in Azerbaijan started in 1993 after the adoption of the Law on the Privatization of State Property. The Law launched the first privatization programme that addressed small and some medium-sized enterprises. The second programme was launched in 2000 with the adoption of a new Law on the Privatization of State Property supplemented by the Second State Programme on the Privatization of State Property adopted by Presidential Decree 383 of

10 August 2000. It covers medium-sized and   large enterprises as well as whole industrial sectors.

Table 2.7: Total environment-related assistance, 1996-2001

(million manats)

1996	1997	1998	1999	2000	2001	Total
180	..	8,489	2,545	104,593	14,384	130,191

Source : OECD. Financing Environmental Protection in Eastern Europe, Caucasus and Central Asia (EECCA). Background Report, 2003.

Notes : Data for 2001 are preliminary. TACIS data for 1998-99 are not available.

By 1 January 2003, some 28,000 small and 1,200 medium-sized enterprises had been privatized, and 1,465 stock companies had been established.

According to the 2000 Law on the Privatization of State Property (art. 6), underground resources, forests, water resources, protected nature areas and the Azerbaijani sector of the Caspian Sea are not subject to privatization. The Law also establishes environmental requirements for privatization. These oblige the seller to provide the buyer with information on the environmental conditions of the privatized object. The Law also foresees the incorporation of specific environmental requirements in sales agreements.

Presidential Decree 383 established, inter alia, that for the sale of State property through investment tenders, bidders should propose investment programmes that cover environmental investments, among other things. The Cabinet of Ministers should define the environmental requirements of privatization procedures. The Decree stipulated, furthermore, that information about privatized enterprises provided to potential bidders should include information on environmental conditions.

Presidential Decree 533 Approving the Rules for Investment Tenders of 17 June 2001 obliged bidders to offer environmental improvements, among other things.

There are cases where privatization actually led to environmental improvements, for example in the Baku tobacco factory, Baku steel and the Garadag cement plant, which modernized and cut their emissions. However, the new owners of privatized enterprisers are concerned first and foremost with improving their profits and seem to ignore even minimal environmental requirements. For instance, the Sumgayit aluminium plant managed to avoid mandatory inspections by the regional environmental committee for at least six months.

There also seems to be a problem with the environmental conditions stipulated in sales contracts for foreign investors. To attract bidders, the Department of Management and Privatisation of State Property of the Ministry of Economic Development tends to weaken environmental requirements. It writes off liabilities for past environmental damage and for environmental damage that continues to be caused by the obsolete equipment of the enterprise before its replacement after the implementation of the investment programme (that takes on average some two years). It also offers grace periods for winning bidders of investment projects regarding the compliance with emission limits and payment of emission charges. At the same time, regional environmental committees are imposing strict compliance with the applicable environmental legislation on all enterprises without exception. The current contradictions of the privatization process may be explained by the fact that the Ministry of Ecology and Natural Resources does not participate in the process.

2.5   Policy objectives and management

The policy framework

The Law on Nature Protection and Environmental Management of 25 February 1992 established the overall objectives for environmental policy and a system of economic instruments. The 1999 Law on Environmental Protection foresees the introduction of new types of economic instruments, including environmental subsidies, environmental insurance and environmental auditing, as well as penalties to promote compliance with environmental requirements.

The 1998 National Environmental Action Plan envisages:

• The introduction of an inflation-adjusted system of penalties and fees for public and private enterprises, to regulate compliance with environmental standards;

- An increase in the funds allocated for environmental management and protection through an environmental fund;
- Amending environmental laws to define liability for past pollution during privatization.

Legal framework

The most significant legislation for economic instruments in environmental protection and natural resource management is:

- The Law on Nature Protection and Environmental Management of 25 February 1992
- The Law on the Privatization of State Property of 1993
- The Law on Environmental Protection of 1999
- The Law on the Privatization of State Property of 16 May 2000
- The Presidential Decree of 10 August 2000 Approving the Second State Programme on the Privatization of State Property
- The Presidential Decree of 17 June 2001 Approving the Rules for Investment Tenders
- Resolution 122 of 3 March 1992 of the Cabinet of Ministers on the Introduction of payments for the use of natural resources, of charges for the emission of pollutants into the environment and on the use of the funds raised from these payments and charges
- Resolution 319 of 8 June 1992 of the Cabinet of Ministers Approving the payment rates for environmental damage caused by the illegal use of mineral resources
- Resolution 216 of 1 May 1993 of the Cabinet of Ministers Amending its Resolutions 122 of 3 March 1992 and 319 of 8 June 1992
- Resolution 239 of 29 December 1998 of the Cabinet of Ministers Amending Resolution 319 of 8 June 1992
- Resolution 41 of 22 February 2001 of the Cabinet of Ministers Approving the Regulations of the State Environmental Protection Fund, Rules for Conducting Environmental Auditing Activities and the Rules for Financing State Environmental Expertise.

Institutional framework

Several governmental bodies are involved in the preparation, formulation and implementation of economic instruments to control environmental

pollution and the mismanagement of natural resources.

The Ministry of Ecology and Natural Resources drafts legislation and regulations setting the scope of application of economic instruments in the fields of air, land and water pollution, waste disposal, the protection of flora (including forests) and fauna (including fish) and the extraction of minerals (with the exception of oil and gas). Regional environmental committees, nature reserve inspectors, the State Environmental Inspectorate, the Forestry Development Department and the Fisheries Department collect the environmental charges, fees, fines and compensation for environmental damage.

The State Committee for Land and Cartography determines the rates of the land-use tax.

The Ministry of Taxes collects the land tax from legal persons and other environment-related taxes.

The Ministry of Economic Development regulates the prices of electricity, gas, crude oil and oil products including petrol, and water supply for specific uses. It also oversees and operates the privatization process.

The executive powers and municipalities regulate heating and municipal waste disposal tariffs. They collect municipal waste disposal charges and land taxes from physical persons.

Local water-supply agencies (*vodocanals*) collect water abstraction fees.

2.6   Conclusions and recommendations

Azerbaijan has faced severe public sector budget constraints throughout the decade. These have resulted from, among other things, the fall in national income compared to the pre-transition period. This has reduced the availability of public finance for all socially important purposes, including the environment. Inadequate governmental funding, which remains the key source of finance for environmental protection, is a major obstacle for the attainment of environmental policy objectives.

Preparation of the budget of the Ministry of Ecology and Natural Resources is implemented with the participation of relevant departments and other bodies. Over the last two years, the Ministry has succeeded in increasing the funds by

150 percent, and these expenditures have been

primarily targeted to strengthening and renewing technical capacities. The process is open and transparent, but funds are inadequate and could be used with greater efficiency, transparency and accountability.

Recommendation 2.1:
The Ministry of Ecology and Natural Resources should improve the management of the State Environmental Protection Fund by addressing its accountability, transparency, cost-effectiveness and environmental effectiveness. The creation of an advisory board for the Fund with the participation of all interested parties, including the environmental NGO community, should be considered

While private and corporate resources (including the banking, finance and investment sectors) represent a valuable potential source of financing, their capacity is still used insufficiently. There is a need to harness commercial and foreign sources of financing for environmental investments. At present, however, there are cases where environmental requirements have been weakened in sales contracts for foreign investments. It is essential that the Ministry of Ecology and Natural Resources be involved in decision-making in the privatization process.

Recommendation 2.2:
(a) The Ministry of Ecology and Natural Resources jointly with the Ministries of Economic Development, of Taxes and of Finance should:
  - *Develop incentives for the public sector to effectively leverage private and foreign finance for the environment; and*
  - *Build the capacity of the executive powers and municipalities to prepare environmental projects that can be co-financed on commercial terms.*
(b) The Ministry of Ecology and Natural Resources should be involved in the decision-making in the privatization process to promote environmental investments by the new enterprise owners.

Azerbaijan has not fully explored the possibilities for attracting foreign assistance for environmental improvements. Assistance is ad hoc and there is no policy or programme to raise external funds. Azerbaijan is eligible, however, for debt relief as part of the initiative on debt relief for the poorest countries of the former Soviet Union. The potential

for debt-for-nature/environment swap initiatives should be discussed between the Ministry of Ecology and Natural Resources and the Ministry of Finance.

Recommendation 2.3:
(a) The Ministry of Ecology and Natural Resources should develop a project portfolio aimed at solving priority environmental problems for submission to prospective donors. Projects should link environmental objectives with poverty reduction, local social and economic development, and strengthening governance. Beneficiaries should be directly involved in both project negotiation and implementation.
(b) The Ministry of Ecology and Natural Resources should also enter into discussions with the Ministry of Finance to prepare expenditure programmes, aimed at solving specific environmental problems, which are not only national but are above all international priorities (e.g. global or transboundary environmental problems).

Azerbaijan is richly endowed in natural resources. Despite their importance for the economy, natural resources do not contribute as much as they could to public revenues. Failure to appropriately tax and reinvest resource rent means that Azerbaijan perpetuates the tradition of exporting low-value-added raw materials and excessive capital outflows. Electricity, heating, water supply and waste disposal tariffs are low and users are not encouraged to save natural resources.

Recommendation 2.4:
The Cabinet of Ministers should proceed with the gradual elimination of environmentally harmful subsidies, starting with the energy sector using the UNECE Guidelines on Reforming Energy Pricing and Subsidies, which were endorsed at the Kiev Ministerial Conference "Environment for Europe" in 2003. The executive powers and municipalities should improve the collection of payments for water consumption and for municipal waste collection and disposal. This measure should be accompanied with a stepwise increase in tariffs to make the respective services self-financing.

Azerbaijan has introduced a wide range of environmental charges and other environment-related economic instruments. They are not, however, promoting changes in behaviour of businesses to prevent or reduce environmental pollution. The generally low rates, the failure to enforce the legislation and frequently inefficient collection are all factors that have weakened the efficiency and environmental impact of the system. The aggregated revenue-raising capacity of pollution charges and other environmental economic instruments is too small to create a critical mass of resources to support significant environmental improvements. The revenue from these instruments represents only some 0.02% of the revenue from all types of charges and taxes in the country. Consequently, Azerbaijan needs to revise its policy of using economic instruments for environmental protection and the management of natural resources.

Recommendation 2.5:
The Ministry of Ecology and Natural Resources should initiate a reform of environmental charges, fees, fines and compensation. This should involve, in particular, raising relevant rates to a level that would provide incentives to prevent or reduce pollution and the misuse of natural resources, and increase revenue substantively.

Chapter 3

ENVIRONMENTAL INFORMATION AND PUBLIC PARTICIPATION

3.1   Introduction

Over the past decade, environmental monitoring in Azerbaijan has been reduced in the face of severe economic conditions. However, in the last two years the Ministry of Ecology and Natural Resources has taken measures to improve the situation. It is making efforts to maintain monitoring in key areas such as air and water quality, and pollution of the Caspian Sea. Azerbaijan is consolidating environmental monitoring by improving data exchange, processing, publication and storage. Significant work remains, however, including the replacement of ageing equipment for sampling and measurement, and the further development of a coordinated national network. Regulations on environmental monitoring have recently been submitted to the Cabinet of Ministers for approval.

Specific monitoring activities need particular attention in Azerbaijan. Soil, biodiversity and waste monitoring are weak. Industrial emissions are not well monitored, reducing the effectiveness of policy instruments such as emissions charges and fines. Monitoring of transboundary air and water pollution needs strengthening. Outdated equipment is used to analyse air, water and soil samples.

Environmental reporting encompasses the various "outputs" of monitoring and information systems. This is a key area that requires attention in Azerbaijan. Better information is needed to support national policy. Azerbaijan does not publish a regular, national state-of-the-environment report. The last report was prepared in 1996. The public has an access to environmental information through the recently established web site of the Ministry of Ecology and Natural Resources.

3.2   Environmental monitoring and data management

Monitoring activities

The National Department of Environmental Monitoring (NDEM) within the Ministry of Ecology and Natural Resources operates 26 air-pollution monitoring stations in eight cities: Baku, Ganja, Sumgayit, Mingechevir, Ali Bayramly, Lenkeran, Sheky and Nakhchivan. Eighteen parameters are measured. The Department carries out spot checks at polluting sites in Baku. Precipitation is measured at 19 monitoring stations and 8 parameters are analysed. There are 29 stations located in 30 bodies of water in the country that measure surface water pollution. Measurements cover basic ions, gases, nutrients, pollutants, and hydrological and physical conditions. The Caspian Complex Environmental Monitoring Department monitors the pollution of the Caspian Sea from the shore and incoming rivers, and shares these results with NDEM.

NDEM regularly measures 17 soil quality parameters at 19 stations in different regions and industrial sites. Thirty monitoring stations measure the level of background radiation and 11 stations measure the radioactivity of atmospheric aerosols. The expedition group within the Department takes air, water and soil samples on the spot at a short notice in environmental emergencies. It also takes samples sporadically to check the sampling results of regional monitoring networks. A permanent Gabala Expedition was created in late 2002 to monitor sources of electromagnetic pollution in the area and their environmental impact.

NDEM prepared an inventory of the bird population in five country regions. NDEM does not collect data on waste and material flows. Specific waste data are reported by enterprises and the executive powers of major cities to the State Statistical Committee.

No countrywide inventory of pollution sources has been prepared in Azerbaijan. The Caspian Complex Environmental Monitoring Department compiled an inventory of pollution sources along the 825 km coastline of the Caspian Sea. The inventory covers 259 watercourses flowing into the sea. It helped to identify 15 rivers of concern for the marine environment and to design monitoring and pollution control measures. Under the assistance programme Jaika (Koku-saiko-ku) of the Japanese Government, an inventory of pollution sources in Absheron was prepared in 2000, but the results are not used for monitoring (or regulatory) purposes.

Monitoring methods, classifications, standards and protocols inherited from the former Soviet Union and those adopted within the Commonwealth of Independent States are generally applied in Azerbaijan. Only guidelines on air pollution measurement were prepared in the national language. Owing to an overall lack of methodological coordination, the results obtained from monitoring networks operated by NDEM, regional environmental committees, the National Hydrometeorological Department and the Sanitary Epidemiological Inspectorate of the Ministry of Health are frequently incomparable and not complementary. There is no centralized or local electronic network for transmission of monitoring data.

The NDEM Central Analytical Laboratory analyses samples taken from the monitoring stations or during spot checks. Few regional environmental committees have their own laboratories to analyse samples taken during inspections of industrial discharges. Other committees submit their monitoring samples to the Central Analytical Laboratory. There is no coordination between the central and regional laboratories, and the latter lack methodological guidance as they do not report to NDEM. At the same time, 15 regional hydrometeorological laboratories under the National Hydrometeorological Department and a number of sanitary and epidemiological laboratories under the Ministry of Health that analyse air, water, soil and radiation samples from the respective independent networks continue to operate.

The budget of NDEM (without the Central Analytical Laboratory) amounted to 340,338,000 manats in 2002 and to 422,229,000 manats (some US$ 77,000) in 2003. Almost no expenditures were earmarked in the budget for the procurement of monitoring equipment. Some 60% of the budget allocations go to its 148 staff, whose monthly salaries vary from 117,000 to 400,000 manats (roughly US$ 21 to US$ 72).

The economic difficulties that Azerbaijan has experienced for many years have led to the degradation of measuring and analytical equipment at monitoring stations and laboratories. This severely restricts the required monitoring activities. For instance, owing to the lack of resources, only 71% of the planned air pollution samples, 45% of the planned precipitation samples and 62% of the planned surface water pollution samples were taken and analysed in the first half of 2003. As a result, the necessary continuity of measurements was interrupted at a number of monitoring stations. The situation with only a few remaining regional analytical laboratories (12 existed in the past) is alarming as they lack technical devices and the means to pursue analysis. The immediate minimal requirements as assessed by the Ministry amount to US$ 200,000 for equipment and chemicals. The current expenditures are covered by the State budget only. Polluting enterprises do not share environmental monitoring costs.

Environmental publications and databases

The Ministry of Ecology and Natural Resources publishes five regular bulletins with monitoring results. Four bulletins, on hydrometeorological conditions, the state of the environment, Caspian Sea pollution and seismic and geodynamic conditions, are published daily. They contain current information and forecasts on, in particular, river and reservoir conditions, expected water availability, air quality and pollution in seven major cities, background radiation in seven regions, pollution of the Caspian Sea, radon concentrations in soil and the chemical condition of groundwater. A further state-of-the-environment bulletin that is published monthly provides information, in particular, on hydrometeorological conditions, the results of pollution monitoring in the Caspian Sea and the activities of the Ministry of Ecology and Natural Resources. These bulletins are circulated within the Ministry itself and are submitted to the President's administration, the Cabinet of Ministers, Parliament, selected ministries, other public entities and municipal authorities.

The Ministry of Ecology and Natural Resources is establishing a system for producing regular state-of-the-environment reports of Azerbaijan. This involves the creation of an interdepartmental commission and an expert group and deciding on a structure for the report. In doing so, the Ministry is following the Guidelines on the Preparation of Governmental Reports on the State and Protection of the Environment, which were endorsed at the Kiev Ministerial Conference "Environment for Europe" in May 2003. The first report is expected to be published in 2004.

The State Statistical Committee publishes an annual statistical yearbook on the environment. This bilingual (Azeri and English) publication contains statistical data on the population, land resources, forests, the protection and use of water resources, the protection of the atmosphere, waste, geological exploration and energy, environmental expenditures and international comparisons. Enterprises, the executive powers, the Ministry of Ecology and Natural Resources and other relevant central administrations provide input data.

In 2001, the Ministry of Ecology and Natural Resources set up the State information and archive database on environmental protection and the use of natural resources. It stores data and information on the environment, hydrometeorology, forestry and geology. Environmental information covers, mainly, air pollution, the state of surface and groundwater resources, the Caspian Sea, land degradation, the state of nature reserves and biodiversity, environmental damage caused by economic activities, and charges and fines paid by polluters. The hydrometeorological and geological databases together with the environmental monitoring bulletins and monthly and annual reports of the main departments and services of the Ministry provide the basis for the database. Regional environmental committees, the State Committee of Amelioration and Water Management and the State Oil Company also contribute their reports. Sixty-two persons service the database, which in the very near future is expected to expand to collect data on local and international environmental projects and on the use of natural resources by all businesses in the country.

The Environmental Electronic Information Centre, an institutional part of the database management, develops the web site of the Ministry of Ecology and Natural Resources, which was created in August 2002. In April 2003, the web site was redesigned. Its information is primarily in the national language but the English version of the site is slowly expanding. It is updated regularly and the state-of-the-environment bulletin is posted daily. Under the ongoing UNECE/EEA environmental monitoring project supported by TACIS funds, the Centre is developing an inventory of data sources. The objective is to post on the Ministry's web site by the end of 2003 some 150 sources of environmental metadata described in accordance with EEA tools and guidelines.

Prospects

The National Department on Environmental Monitoring prepared draft rules for environmental monitoring and for environmental data collection from ministries and other public authorities and the Ministry has submitted these to the Cabinet of Ministers for adoption.

3.3   Outreach

Communication with the public

The Ministry of Ecology and Natural Resources is making substantive efforts to ensure that environmental information is accessible to the public. On 11 September 2003, an Aarhus Information Centre was opened within the Ministry, with the support of OSCE. The Ministry has also established a web site, and it posts a wide range of information on its own activities on it (see above). In May 2001, it established a press office to improve links with the media and NGOs and to produce information publications for the general public. The press office issues press releases a few times each week and circulates these among journalists and NGOs. It also posts the press releases on the Ministry's web site. The Ministry publishes the *Priroda Azerbaijana* (Nature of Azerbaijan) magazine every other month. It has a circulation of 2,000 copies and is intended for a wide audience.

The press office is developing a concept for a weekly newspaper on the environment to promote, in particular, a dialogue on key issues of public concern such as transport emissions, water problems, unauthorized landfills, untreated water discharges into the sea and reforestation in rural areas. It is considering practical ways and means to produce information leaflets, to support the production of environmental TV clips and environmental advertisements along streets in cities, to organize environmental training for journalists, and to raise funds for environmental awareness activities with foreign companies operating in Azerbaijan.

At the regional level, only the Baku-Absheron environmental committee has an information division that cooperates with mass media.

Environmental NGOs

According to a recent inventory by the Regional Environmental Centre for the Caucasus, there are 85 environmental NGOs in Azerbaijan. The Initiative for Social Action and Renewal in Eurasia (ISAR-Azerbaijan) compiles and regularly updates a bilingual (Azeri and English) directory of environmental NGOs of Azerbaijan with records on both the NGOs themselves and their activities. Many of the NGOs, such as Agro-Ecoconsulting Centre, Azerbaijan Society for the Protection of Animals, Bird Protection Society, Eco-Energy Academy, Ecolex Legal Environmental Centre, Ecosphere Social and Ecological Centre, Fovgal Expert Association on Emergencies and Human Safety, Gulum Children Association, Hezri Association for Development and Environmental Research, Human and Environment, Piligrim and Rusgyar, have gained broad public support for their initiatives in environmental education, environmental law, ecotourism, wildlife protection, environmental technology, waste management or other areas.

So far, only 25 environmental NGOs have formally registered with the Ministry of Justice. It appears that the latter applies stringent registration procedures for all NGOs. This creates difficulties for NGO operations. There is no financial support for NGOs from public authorities, and they are largely dependent on foreign assistance for their projects. According to the national legislation, NGOs are subject to value-added tax (VAT) and other taxation. The 2000 legislation on NGOs does not establish procedures to explicitly regulate NGO operations. All this impedes a more active involvement of civil society in environmental activities in Azerbaijan.

In May 2002, the Ministry of Ecology and Natural Resources started a structured dialogue with the environmental NGO community. The Minister invites representatives of 20-25 NGOs and mass media representatives on a regular basis to round-table discussions on salient issues of the national environmental policy. These have included the environmental implications of the Baku-Ceyhan pipeline, reforestation, environmental inspections, environmental emergencies and modalities of NGO interaction with the Ministry. The Ministry's intention is to form a consultative NGO council at the Ministry. On some occasions the Ministry has invited NGO representatives and journalists to attend meetings of its collegium.

3.4 Access to information and public participation

Practical application

The Law on Access to Environmental Information and the Law on Freedom of Information establish restrictions for public access to environmental information. These restrictions largely follow the relevant provisions of the Aarhus Convention. Nevertheless, the national legislation gives the public authorities the power to interpret these restrictions broadly, and this is widely used on the basis of State security considerations.

The lack of regulations establishing clear procedures and rules impedes the practical implementation of the citizen's right of access to environmental information.

The practical experience in Azerbaijan in providing the public with possibilities for participating in environmental decision-making is limited, to a great extent, to environmental impact assessment procedures (see below). The public does not participate in the development of strategies, programmes and plans of relevance to the environment. It can receive information about them only through the mass media, if at all. Neither are environmental permitting and licensing procedures open to the public, nor is the information on them accessible to the public.

In pursuance of the Law on Environmental Safety, the public may participate in the discussion in the parliament of draft environmental laws prepared by the public authorities. There is some initial practice for involving NGOs in discussing draft environmental regulations that are subject to adoption by the President, the Cabinet of Ministers or by relevant ministries.

National legislation establishes cases when a person can resort to judicial procedures to enforce his or her right of access to environmental information and participate in environmental decision-making. These cases generally follow the relevant provisions of the Aarhus Convention. The lack of precision in legal norms ensuring the protection of individual rights, however, poses difficulties in considering environmental claims in court. Furthermore, as judges are designated by the executive powers and report to the Ministry of Justice, the public is generally not very confident in the national judicial system and people tend to avoid resorting to courts to defend their rights. They seem to prefer to use administrative procedures by appealing to supervisory bodies or officials instead.

Participation in environmental impact assessment (EIA)

According to the Regulation on Environmental Impact Assessment (1996), the project proponent and the State Environmental Impact Assessment Department of the Ministry of Ecology and Natural Resources are obliged to consult the public before preparing the EIA documentation and to organize public hearings once this documentation has been prepared. Representatives of the public have the right to be included in expert groups that examine EIA documentation. The public can submit comments on EIA documentation and on the concluding document by the governmental expert group in various ways as stipulated in the Regulation.

The State Environmental Impact Assessment Department generally follows the provisions of the Regulation on EIA. It was particularly successful in cases where foreign companies proposed economic development projects. In many instances, the public was provided with EIA documentation on proposed activities and open hearings were organized with active public participation. Frequently, big companies, especially foreign ones, have disseminated environmental information on proposed projects and announcements about public hearings via popular newspapers and TV. High company

executives and experts of the Ministry of Ecology and Natural Resources took part in the hearings to brief the public.

Representatives of research and academic institutions and the mass media take an active part in public hearings. The participation of NGOs, including environmental ones, is rather weak. The State Environmental Impact Assessment Department does not generally inform the public about contact points to whom written comments should be sent or about deadlines for comments. Neither does it inform members of the public whether their comments were taken into account in the decisions as a result of EIA and, if not, on what grounds.

3.5 Policy objectives and management

The policy framework

Azerbaijan's National Environmental Action Plan of 1998 envisages, inter alia:

- Critical capital investments in monitoring and laboratory equipment
- Eliminating the overlap in monitoring responsibilities
- New legislation to provide incentives for self-monitoring of pollution by industrial plants.

The 2003 National Programme on Environmentally Sustainable Socio-economic Development foresees increasing the role of NGOs in the implementation of economic and social projects, to raise public awareness by organizing art, photo and poster exhibitions, festivals and contests on environmental protection, and to promote environmental culture and environmental ethics among the general public.

Legal framework

Data collection and reporting

Various legal acts establish priorities for environmental monitoring and requirements for environmental data collection and management. The Law on Environmental Protection (1999) sets up the overall system of State monitoring of the environment and natural resources. The Law on Hydrometeorology (1998) provides for hydrometeorological observations and monitoring of environmental pollution. Legislation on water (1997) and forestry (1997), the Laws on the

Radiation Safety of the Population (1997), on Fisheries (1998), on Air Protection (2001), on the State Land Cadastre, Soil Monitoring and Land Management (1999), on Mineral Resources (1998), on Industrial and Municipal Waste (1998), on Wildlife (1999), on Specially Protected Natural Areas and Objects (2000) and on Pesticides and Agrochemicals (1997) lay down requirements for monitoring in these respective areas. Furthermore, legal acts on the ratification of global and regional environmental conventions contain specific monitoring provisions to ensure compliance with these international legal instruments.

According to article 4 of the Law on Environmental Protection, the public authorities are obliged to establish rules for the collection, processing and supply of environmental information and statistics. The Law on Environmental Safety defines the public authorities that are responsible for the provision of environmental information including monitoring of the environment and natural resources.

Access to information and public participation

Legislation on public access to environmental information, on public participation in environmental decision-making and on access to justice in environmental matters includes the Constitution, the Law on Environmental Protection, the Law on Access to Environmental Information, the Law on Sanitary and Epidemiological Services, the Law on Information, its Dissemination and Protection, the Law on Freedom of Information, the Law on Procedures for the Consideration of Citizen's Appeals, the Law on the Mass Media and the Law on State Secrets. Azerbaijan also ratified the Aarhus Convention on Access to Information, Public Participation in Decision-making and Access to Justice in Environmental Matters, the Espoo Convention on Environmental Impact Assessment in a Transboundary Context and other relevant international treaties. According to the Constitution (chapter 148, part 2, and chapter 152), the requirements of applicable international treaties take precedence over national legislation.

The Laws on Environmental Protection and on Access to Environmental Information establish the scope of environmental information to be provided to the public: the state of the environment and impacts on it, environmental regulations, standards and expenditures. The Law on Sanitary and Epidemiological Services (art. 12) requires public authorities to collect and provide the public with full and reliable information on environmental health. Public access to other types of environmental information that are not specifically mentioned in the national legislation but are covered by article 2, paragraph 3, of the Aarhus Convention, such as plans and programmes of relevance to the environment, is ensured by the national legislation that establishes the direct applicability of provisions of international law.

The Law on Access to Environmental Information (art. 5) obliges the national and local authorities to establish information collection systems including cadastres of natural resources and pollution registers, update continuously environmental databases including in electronic form, provide the public with lists of authorities possessing information and with environmental documentation, and to disseminate information about emergencies without delay. They should publish, at least once every three years, state-of the-environment reports and make these accessible to the public. Furthermore, according to the Law on Procedures for the Consideration of Citizen's Appeals, the public has the right to request environmental information from national and foreign enterprises and organizations operating in Azerbaijan.

National legislation sets time limits for providing the public with environmental information upon request. This should be done within 15-30 days or maximum 2 months in exceptional cases and in written form only. Applicants are not obliged to prove their specific interest in the requested information. The information should be provided either free of charge or upon payment of reasonable costs. Legislation establishes cost-recovery cases for the provision of specific types of information (e.g. hydrometeorological data) as well as appeals procedures for cases of violation of citizens' rights of access to information.

Institutional framework

The Ministry of Ecology and Natural Resources, through its National Department of Environmental Monitoring, the National Hydrometeorological Department, the Laboratory of Integrated Monitoring of the Caspian Sea, the Forestry Development Department and the Department of Bioresources and Specially Protected Areas, collects the bulk of environmental data in the country. The State Information and Archive Database on Environmental Protection and Use of

Natural Resources stores these data. Other institutions dealing with the collection and management of environmental data include the State Statistical Committee, the Sanitary Epidemiological Inspectorate of the Ministry of Health and the State Committee of Amelioration and Water Management.

The National Department of Environmental Monitoring (NDEM), established in 2002, aims at creating an integrated environmental monitoring system in Azerbaijan and assisting decision makers in the Ministry of Ecology and Natural Resources to prevent, control and reduce adverse environmental impacts, and to assess the effectiveness of policies and measures. Its tasks are:

- To collect, process and assess data on the state of the environment, on the causes of ongoing or expected changes in the environment, on critical loads and levels as well as chemical, physical and social impacts on the environment, and on the carrying capacity of the environment;
- To assess the actual state of the environment (air, surface water and groundwater, soils, coastal areas, vegetation including forests, and fauna) and pollution by specific substances;
- To forecast the state of the environment according to the expected pollution levels and the use of environmental resources.

3.6 Conclusions and Recommendations

The objectives of the National Environmental Action Plan of 1998 to strengthen environmental monitoring have not yet been met. In spite of some recent consolidation of monitoring activities within the Ministry of Ecology and Natural Resources, several administrations continue independent monitoring using poorly equipped analytical laboratories. The National Department of Environmental Monitoring (NDEM) within the Ministry of Ecology and Natural Resources operates 26 air-pollution monitoring stations in eight cities, and the Caspian Complex Environmental Monitoring Department monitors the pollution of the Caspian Sea. At the same time, 15 regional hydrometeorological laboratories under the National Hydrometeorological Department and a number of sanitary and epidemiological laboratories under the Ministry of Health continue to analyse air, water, and soil and radiation samples from the respective independent networks.

Recommendation 3.1:
The Ministry of Ecology and Natural Resources should consolidate further the role of its National Department of Environmental Monitoring as lead environmental monitoring agency responsible for core monitoring activities and coordination with all other administrations, research institutes, regional environmental centres and NGOs, collecting and processing environmental data. A merger of the Caspian Complex Environmental Monitoring Department and the National Department for Environmental Monitoring (NDEM), and the transfer of hydrometeorological laboratories to NDEM could be considered, among other measures.

Besides the Ministry of Ecology and Natural Resources, several other ministries and agencies deal with the collection and management of environmental data, including the Sanitary Epidemiological Services under the Ministry of Health, the State Committee of Amelioration and Water Management, and the State Statistical Committee. There is practically no coordination among these public authorities on the location of monitoring stations, sampling or data exchange. Apart from the environmental statistical yearbook, the Ministry of Ecology and Natural Resources prepares a number of media- and topic-specific bulletins that are distributed among all ministries and made available in the Aarhus Information Centre in the Ministry. However, at present, there is no comprehensive environmental publication published in Azerbaijan on a regular basis.

Recommendation 3.2:
(a) The Cabinet of Ministers should establish an institutional structure for inter-ministerial cooperation and coordination on environmental monitoring and information with the Ministry of Ecology and Natural Resources having the leading role.
(b) The development of a State system of integrated environmental monitoring and the preparation of a regular governmental report on the state and the protection of the environment should be core responsibilities of this structure (commission), which should be supported by a network of experts responsible for specific monitoring and information activities.

Significant resources are required to replace obsolete monitoring and laboratory equipment for sampling, measurement and analysis. Furthermore, the monitoring of soils, biodiversity, solid and hazardous waste, and some other areas is weak.

Recommendation 3.3:
(a) *The Ministry of Ecology and Natural Resources, when finalizing the State programme for strengthening environmental monitoring for submission to the Cabinet of Ministers, should include a detailed assessment (including cost assessment) of the investment requirements in basic environmental monitoring infrastructure, in particular in raw data collection, analytical and processing capacities, and equipment.*
(b) *The programme should also establish a clear perspective of extending monitoring activities, step-by-step, to soil, waste, biodiversity, and chemicals in ecosystems and foodstuffs to ensure integrated data collection covering quality, quantity, biodiversity and ecosystem aspects from the outset.*

Most enterprises do not monitor their emissions into the environment. Neither do they participate in sharing monitoring costs with environmental authorities. Under the current severe budget restrictions and the lack of foreign assistance, environmental monitoring networks in Azerbaijan are not sustainable.

Recommendation 3.4:
The Ministry of Ecology and Natural Resources should draft legislation making polluting enterprises responsible for monitoring their emissions and waste flows. It should also provide companies with guidance and incentives for voluntary reporting on their environmental performance.

The Ministry of Ecology and Natural Resources publishes environmental bulletins with monitoring results. These bulletins, as well as environment-related bulletins produced by other ministries and agencies, are circulated internally and submitted to the President's administration, the Cabinet of Ministers, Parliament, other public entities and municipal authorities. These materials are made available to the public in the Aarhus Information Centre in the Ministry, and there is a plan to post them on the Ministry's Web site as well. At the same time, the procedure for making them available could be further improved to make them more accessible to the general public.

Recommendation 3.5:
The Ministry of Ecology and Natural Resources, the State Statistical Committee, the Ministry of Health and the State Committee of Amelioration and Water Management should make

environmental data, including environmental health data, collected with public funds freely available. They should make every effort to raise external funds, if necessary, to produce compact, easy-to-read products such as booklets presenting key environmental data, indicator reports and thematic leaflets or brochures, and to make them available on the Internet.

The lack of regulations establishing clear procedures and rules impedes the practical implementation of citizens' right of access to environmental information. Public participation in environmental decision-making is limited to participation in EIA procedures. The lack of precision in legal norms ensuring the protection of individual rights poses difficulties in considering environmental claims in court. Stringent registration procedures for NGOs impede a more active involvement of civil society in environmental activities.

Recommendation 3.6:
(a) *The Cabinet of Ministers should issue regulations supplementing existing laws to ensure that unambiguous and detailed procedures are in place guaranteeing public assess to environmental information, public participation and access to justice on environmental issues to comply fully with the Aarhus Convention.*
(b) *These regulations should also simplify the registration procedure for environmental NGOs*

The Ministry of Ecology and Natural Resources has started a dialogue with the environmental NGO community. There are currently regular meetings between the Ministry and NGOs to discuss environmental projects and problems as well as EIA procedures. NGOs are also involved in working parties under the conventions and all projects run by the Ministry include NGOs as well. The dialogue between the Ministry and NGOs, however, could be better structured and NGOs should be given greater opportunities for contributing to national environmental policy-making.

Recommendation 3.7:
The Ministry of Ecology and Natural Resources should establish a consultative body and procedure at the ministry with broad participation of national environmental NGOs in the development of environmental legislation, programmes and plans.

Chapter 4

INTERNATIONAL COOPERATION

4.1 General objectives for international cooperation

One year after its date of independence, Azerbaijan joined the United Nations in 1992. It began the process of ratifying several global and regional conventions, and has taken an active part in European institutions such as the Organization for Security and Co-operation in Europe (OSCE) and the Council of Europe.

Azerbaijan cooperates with a number of United Nations programmes and specialized agencies, such as the United Nations Environment Programme (UNEP), the United Nations Development Programme (UNDP), the United Nations Educational, Scientific and Cultural Organization (UNESCO), the World Meteorological Organization (WMO), the World Health Organization (WHO), and the United Nations Industrial Development Organization (UNIDO). It is currently a member of the United Nations Commission on Sustainable Development and the Economic and Social Council.

For geographical and historical reasons, Azerbaijan is a member State of both the United Nations Economic Commission for Europe (UNECE) (since July 1993) and the United Nations Economic and Social Commission for Asia and the Pacific (ESCAP) (since July 1992).

The country's international cooperation on environmental protection has been influenced by its complex geopolitical situation. The country is located in the South Caucasus, which is an important crossroad linking European, Central Asian and Western Asian countries and therefore has a strategic geographic position.

The shortage of drinking water and other water resources, the contamination of both surface water and groundwater, the high degree of land salinization and soil degradation, air pollution, waste generation, particularly toxic and radioactive waste, and the loss of biological diversity are among its most significant environmental problems.

The scale and complexity of these problems need an integrated and multifaceted approach and significant financial resources. In this context, international cooperation plays an important role in providing access to international investments, international experience and practice and in introducing cleaner technologies and processes.

At present, the main policy objectives in international cooperation are: participating in the international legislative and environmental policy process and cooperating with international organizations; implementing modern international environmental policies, standards and norms, and technology; attracting foreign assistance for environmental protection, nuclear safety and the rational use of natural resources; solving problems connected to transboundary waters, such as the Kura, Araz and Samur rivers, or shared waters (Caspian Sea); bilateral cooperation with neighbouring countries, other important partner and donor countries; and participating in regional nature conservation for the Caspian Sea and natural parks.

The following priorities for cooperation are particularly relevant in the regional context:

- Harmonization of legislation in accordance with the EU directives (See chapter 1 on Policy, legal and institutional framework);
- Drawing-up of standards and norms to meet EU and international standards;
- Participation in the "Environment for Europe" process;
- Caspian Sea protection; and
- Bilateral cooperation with neighbouring countries, including Georgia, the Islamic Republic of Iran, Kazakhstan, the Russian Federation, Turkmenistan and Turkey.

The institutions generally involved in international environmental cooperation are the Department of International Cooperation and, to some extent, the Department of Environmental Policy and Protection within the Ministry of Ecology and Natural Resources. Other parts of the government can also be directly involved in international cooperation for the environment, such as the Cabinet of Ministers, the Presidential Administration, and the Ministry of Foreign Affairs. Each ministry has a department of international cooperation. However, the Ministry of Ecology and Natural Resources coordinates projects related to the environment.

Pursuant to the Vienna Convention on the Law of Treaties, in 1995 Azerbaijan adopted the Law on Signing, Implementing and Withdrawing from International Agreements, No. 1057, which sets out the steps at a national level for the ratification of multilateral environment agreements (MEAs).

Azerbaijan is a Party to some 17 global and regional conventions or agreements on environmental protection and nuclear safety, and to six related protocols (see annex II). Moreover, the Stockholm Convention on Persistent Organic Pollutants (POPs) and the Bonn Convention on the Conservation of Migratory Species of Wild Animals, have been submitted to Parliament for ratification. The Ministry of Ecology and Natural Resources started the process for the ratification of the Agreement on the Conservation of African-Eurasian Migratory Waterbirds, and a number of other conventions and agreements are also under consideration.

Azerbaijan often finds it difficult to comply with and enforce the multilateral environmental agreements that it has already ratified because of a lack of financial commitment, institutional and human capacity and experience. This is compounded by the fact that the Ministry, which was only established in May 2001, is new and is still being reorganized.

Despite compliance and enforcement problems, Azerbaijan nevertheless considers the signing and ratification of multilateral environmental agreements to be a priority, because this provides incentives for harmonizing national legislation with international practices. When it ratifies such agreements, a national law that incorporates or transposes the international agreement is adopted as part of the ratification. If its national norms are stricter than the international standards, the national norms prevail.

4.2   Global cooperation

After its independence, Azerbaijan also became a member of various international financial institutions. Azerbaijan joined the World Bank in September 1992 and the Islamic Development Bank also in the same year. It has been a member of the Global Environment Facility (GEF) since July 1995, represented by Switzerland on the GEF Council. In 1999, Azerbaijan became a member of the Asian Development Bank (ADB). In a continually improving investment environment, the European Bank for Reconstruction and Development (EBRD) has been assisting Azerbaijan in developing its oil and gas sector with due regard to the environment and transparent management of these resources.

Azerbaijan, together with ten other member States (Albania, Armenia, Bulgaria, Georgia, Greece, Republic of Moldova, Romania, Russian Federation, Turkey and Ukraine) of the Organization of the Black Sea Economic Cooperation (BSEC), established the Black Sea Trade and Development Bank (BSTDB) in 1998 as a regional multilateral financial institution. Its mandate is to accelerate development and promote cooperation among its shareholder countries. It supports regional trade and investment and provides financing for commercial transactions and projects in order to help its member States establish stronger economic links. But as a major component of its objectives, BSTDB promotes environmental sustainability and cleaner technologies in the projects that it finances.

International cooperation and the National Environmental Action Plan

Azerbaijan's National Environmental Action Plan (NEAP) was finalized in 1998 with the support of the World Bank. It defines the key environmental problems (see chapter 1 on Policy, legal and institutional framework).

To address the most urgent actions identified in the National Environmental Action Plan, the Government of Azerbaijan and the World Bank agreed to prepare an urgent environmental investment project. Its development objectives are: (a) restoring the capacity to produce sturgeon fingerlings by building a new sturgeon hatchery near Neftchala; (b) demonstrating mercury clean-up technologies and procedures by cleaning up one area heavily polluted by mercury; (c) testing and demonstrating onshore oil field clean-up methodologies by cleaning up one oil field on the Absheron Peninsula; and (d) improving the institutional and regulatory capacity of the environmental management system.

The project began at the end of 1998 and is to be implemented over five years. It contains three site-specific components and two institution-building components. The latter have already resulted in the establishment of the Project Implementation Unit (PIU or fifth component). The three site-specific components are in the finalization stage. Since the project will end soon, PIU is looking for a sustainable use of the sturgeon hatchery and it wants to use the experience gained from the pilot oil field clean-up project at all oil fields.

Another World Bank project, Shah-Dag Rural Environment Programme, is being implemented to support integrated ecosystem management in the Shah-Dag area, including investments in forest, range and protected area management, rural energy supply and conservation, capacity building, and project management (see Box 9.1 in Chapter 9 on Biodiversity and forest management). The proposed park and buffer zone form a catchment area, which serves as a conservation unit containing essential elements of the Caucasus mountain ecosystem. The main objectives are to promote in rural areas the use of environmentally sound practices in forestry, grazing and energy consumption and to conserve biodiversity, reverse land degradation and improve energy supply and efficiency in the Shah-Dag area of the Azerbaijan Caucasus Mountains (see Chapter 10 on Land use, agriculture and desertification).

Implementation of Agenda 21

In 1992, at the United Nations Conference on Environment and Development (Rio de Janeiro, Brazil), 172 Governments agreed on an ambitious plan of global action entitled Agenda 21. Over the past decade, Azerbaijan has attempted to implement Agenda 21 to the extent possible, and it has made its reports on implementation publicly available. In 2002, it prepared its National Report on Sustainable Development in the Context of the Rio+10 Process on the implementation of Agenda 21, which it submitted to the United Nations Commission on Sustainable Development for the World Summit on Sustainable Development (Johannesburg, South Africa, 2002).

Protection of the atmosphere and prevention of climate change

The closure of a number of heavily polluting industrial enterprises since 1991 has led to a substantial reduction in air pollutants. In 1990 the volume of pollutants was approximately 2.1 million tons, but in 1998 it had declined to 352,000 tons. A number of problems remain, however. Most air filters at the working enterprises are in poor condition, and air pollutants from motor vehicles are on the increase (see Chapter 5 on Air management and transport). To help resolve worldwide air pollution problems, Azerbaijan ratified the United Nations Framework Convention on Climate Change in May 1995 and the Vienna Convention for the Protection of the Ozone Layer in 1996.

In 2000, Azerbaijan ratified the Kyoto Protocol to the United Nations Framework Convention on Climate Change. As a Party to the Convention that is not listed in Annex 1, Azerbaijan has undertaken to develop, implement and disseminate national and regional programmes to reduce the expected impacts of climate change. Non-Annex 1 countries can host clean development mechanism projects. These must be independently certified, giving the countries certified emissions reductions that can be banked from the year 2000, before the 2008-2012 commitment period.

In April 1997, the Government of Azerbaijan issued Decree 560 establishing the State Commission on Climate Change, which included the heads of 18 ministries, committees and firms. One of the first steps taken to comply with articles 4 and 12 of the Convention was the setting-up of a project for Azerbaijan's First National Communication, with financial assistance from GEF and the United Nations Development Programme (UNDP). Its main purpose is to draw up a national plan of action to minimize the negative impact of climate change on the country's economy and the health of its population, and to inform the international community about this. Work on this project has proceeded to:

- Prepare a national inventory of greenhouse gases;
- Assess opportunities for reducing the use of greenhouse gases and draw up proposals to implement a national policy in this regard; and
- Evaluate the impact of climate change on ecosystems and major sectors of the economy, and prepare adaptation measures to minimize losses resulting from such an impact.

Together with 11 other States (Albania, Armenia, Croatia, Georgia, Mongolia, Republic of Moldova, Slovenia, Tajikistan, the former Yugoslav Republic of Macedonia, Turkmenistan and Uzbekistan), Azerbaijan participates in the regional GEF project "Capacity-building for Improving the Quality of Greenhouse Gas Inventories." The project aims to develop capacity for these 12 States to improve emission factors and data collection procedures, and to strengthen national institutions to estimate national greenhouse gas emissions and removals. It is expected to run from June 2002 to June 2005.

Protection of the ozone layer

When it ratified the *Vienna Convention for the Protection of the Ozone Layer* in 1996, Azerbaijan also ratified the *Montreal Protocol on Substances that Deplete the Ozone Layer* and its amendments (*London, Copenhagen and Montreal Amendments*) on phasing out the production and use of ozone-depleting substances (ODS).

A working group was created and, in 1997, United Nations experts conducted special investigations on the production and consumption of ODS. They found that Azerbaijan produced a total of 965.7 tons on ODS (459.4 tons of chlorofluorocarbons (CFCs), 501.1 tons of halons, 0.1 tons of methylbromide, 5.1 tons of hydrochlorofluorocarbons (HCFCs) and 0.5 tons of methylchloroform). With the help of a GEF grant worth US$ 7 million, the Azerbaijan Ozone Centre was created and the Government adopted a strategy to phase out ODS.

Transboundary movement of hazardous waste

Azerbaijan ratified the 1989 *Basel Convention on the Control of Transboundary Movements of Hazardous Wastes and their Disposal* in June 2001. The laws regulating the Basel Convention are the 1998 Law on Industrial and Municipal Waste, the 1999 Law amending

some legislative acts with regard to the application of the Law on Industrial and Municipal Waste, and the 2001 Law on Joining the Basel Convention on the Control of Transboundary Movements of Hazardous Wastes and their Disposal.

In July 2003 the Ministry of Ecology and Natural Resources was appointed as the competent authority for implementation of the Basel Convention (Decree of Cabinet of Ministers 17/17/2015-22, July 2003). The Hazardous Waste Management Agency was appointed as the coordinating body for the implementation of the Convention. A new law on hazardous waste exports and imports, which would be based on the Basel Convention, is in the pipeline. The Agency has already developed a hazardous waste passport, and an inventory and classification of hazardous waste based on the provisions of the Basel Convention are planned. At present data on exports and imports are not available. In the future, customs should provide such information to the Ministry (see Chapter 6 on Management of waste and contaminated sites).

Convention on Biological Diversity and other conventions related to nature protection

Azerbaijan is concerned about the degradation and loss of biodiversity. It has signed up to a number of international agreements protecting biological diversity, wetlands, wild nature and the environment.

Following its approval of the *Convention on Biological Diversity* in 2000, Azerbaijan undertook a number of activities to comply with the Convention. It established the State Committee for Biodiversity and Genetic Resources in 2001 and, with assistance from GEF and UNDP, initiated a project to develop a biodiversity strategy, action plan and national report in May 2002. The project collected and assessed information, and identified the options available to meet the Convention's objectives and requirements. The coming reports will help Azerbaijan fulfil its commitments under the Convention.

As a participant in the Pan-European Biological and Landscape Diversity Strategy, Azerbaijan has developed programmes and plans of action covering different parts of the Strategy. It has also drawn up plans for the development of specially protected territories, the protection of unique and endangered species of flora and fauna, the restoration of wetlands, and the development of mountain ecosystems.

Many species of animals and plants found in Azerbaijan are endemic and do not exist anywhere else in the world. According to the Red Book, hundreds of species of plants and animals are endangered and have been assigned various types of protection status in Azerbaijan. The protection of biodiversity, including migrating birds, is of national, regional and global importance.

Azerbaijan acceded to the *Ramsar Convention on Wetlands of International Importance Especially as Waterfowl Habitat*, the Paris Protocol amending the Convention and the Regina Amendments to Articles 6 and 7 of the Convention (*Regina, Canada, 1987*), in September 1991, and recognizes responsibility for the conservation of 22 water and wetland areas of international importance (total area 678,000 ha). Following the ratification of the Ramsar Convention, two wetlands, namely one in the Gizil-Agaj State Reserve and Lesser Gizil-Agaj Sanctuary and the other in the Ag-Gol State Reserve and Ag-Gol State Sanctuary, were put on the Ramsar list (see chapter 9 on Biodiversity and forest management).

The Ministry of Ecology and Natural Resources realizes its responsibility and role in saving unique wetlands and is taking appropriate measures. In partnership with the Succow Foundation of Germany, the Ministry developed a project for saving Ag-Gol Lake and has already started its implementation. In addition, the Ag-Gol State Reserve was extended from 4,000 ha to 15,000 ha with a view to strengthening the protection of the territory and thereby mitigating the anthropogenic impact of local communities. Currently, the Lake's water quality and levels of inflow and outflow are monitored regularly. Similar hydro-technical activities in the Shirvan State Reserve are meant to eliminate the threat of its wetland drying up.

Azerbaijan acceded to the *Convention on the Conservation of European Wildlife and Natural Habitats* in March 2000. Located on the main flyway of migratory birds, Azerbaijan is an ideal place for waterbirds to winter and nest during migration. Along their migration route, these waterbirds have to rely on a number of wetlands as staging posts to rest and feed. Azerbaijan's

wetlands are also breeding and feeding habitats for millions of waterfowl, shore birds and other wildlife. As mentioned above, Azerbaijan is in the process of ratifying the Bonn Convention on the Conservation of Migratory Species of Wild Animals. It is also in the process of ratifying the Agreement on the Conservation of African-Eurasian Migratory Waterbirds.

Azerbaijan ratified the *Convention for the Protection of the World Cultural and Natural Heritage* in December 1998, but at present only one site of cultural importance falls under this Convention. Its secretariat is considering the inclusion of four natural sites.

Furthermore, Azerbaijan ratified the *Washington Convention on International Trade in Endangered Species of Wild Fauna and Flora (CITES)* in November 1998. It should now set up a CITES management authority. The next steps will have to be the drawing-up of instructions and guidelines for customs officers and environmental inspectors, and training these officials. In this context, Azerbaijan especially needs to address the rehabilitation and replenishment of sturgeon stocks. In order to preserve this natural resource, Azerbaijan has formulated a national policy to protect fish. Sturgeon is also a key issue for international cooperation with the littoral countries of the Caspian Sea (see chapter 8, on selected problems of the Caspian Sea).

Convention to Combat Desertification in Countries Experiencing Serious Drought and/or Desertification, Particularly in Africa

Of Azerbaijan's 8.6 million ha, only 4.7 million ha, or 54.8%, are suitable for cultivation. The quality of its soil even on this land has been seriously degraded by erosion, salinization, swamping, chemical pollution and other processes (see chapter 10 on Land use, agriculture and desertification).

Under such circumstances, signing up to the United Nations Convention to Combat Desertification has been significant for Azerbaijan as a means not only of using its own resources but also of having access to technical and financial assistance from foreign countries, and international cooperation, to combat desertification. In 1998, the President signed the Law on the ratification of the Convention (487-1).

Recently, steps have been taken within the framework of the Convention. The National Report was prepared in 2000. Work started on the preparation of a national plan of action for combating desertification. Various ministries and organizations have been involved in its development. To prevent desertification and soil erosion, Azerbaijan should:

- Collect and analyse information on soil erosion and desertification;
- Collect and analyse information on the hydrological and geomorphologic condition of pastures;
- Issue regulations on the protection of pastures;
- Select and assess priority projects in the framework of the plan; and
- Develop a public awareness campaign on the situation of pastures and their efficient use.

4.3   Cooperation in the framework of UNECE conventions

Convention on Long-range Transboundary Air Pollution

Azerbaijan acceded to the Convention on Long-range Transboundary Air Pollution in July 2002. Under the Convention, Azerbaijan has already provided some data on air pollution (see Chapter 5 on Air quality, transport and environment). The next step is to provide the monitoring scheme and improve it, and to establish a new legal framework to protect air.

Convention on the Protection and Use of Transboundary Watercourses and International Lakes

Azerbaijan acceded to the Convention on the Protection and Use of Transboundary Watercourses and International Lakes in August 2000. It ratified the Protocol on Water and Health in January 2003. It also approved the 2000 Ministerial Declaration of The Hague on Water Security in the 21st Century.

Transboundary waters are an important issue for Azerbaijan and its neighbours. Two transboundary rivers flow to Azerbaijan from Turkey through its neighbouring countries – the Kura through Georgia and the Araz along the border with Armenia and the Islamic Republic of Iran. Their basins extend across the territory of Armenia, Azerbaijan, Georgia, the Islamic Republic of Iran and Turkey (see figure 4.1). These rivers discharge into the Caspian Sea. Azerbaijan

shares the Samur river with the Russian Federation (see Chapter 7 on Water Management).

Azerbaijan relies for 95% of its drinking water on surface water. Almost 80% of the population drinks water from the Kura river. Although its water is already polluted when it enters Azerbaijan, Azerbaijan lacks the equipment to purify water from sewage and industrial discharges. To assess the water quality of the Kura river, the North Atlantic Treaty Organisation has launched a programme to monitor the situation in Azerbaijan, Georgia and Turkey. Each country received 30 stations to monitor water quality. A GEF project, entitled "Regional Partnership for Prevention of Transboundary Degradation of the Kura-Araz Rivers," is under evaluation. Its aim is to establish regional cooperation as a key tool for the long-term sustainable management of the two rivers.

Convention on Environmental Impact Assessment in a Transboundary Context

Azerbaijan acceded to the Convention on Environmental Impact Assessment (EIA) in a Transboundary Context in March 1999. EIA procedures are compiled in the 1996 Handbook for the EIA Process and have been developed in accordance with international standards (see Chapter 1 on Policy, legal and institutional framework). It incorporates the main provisions of the Convention, including provisions for public participation. However, environmental norms for EIA in a transboundary context have not yet been developed (sanitary and hygiene norms are being used at present). Nevertheless, Regulations (no law) on Environmental Impact Assessment were adopted in April 2003 and approved by the Cabinet of Ministers.

Convention on the Transboundary Effects of Industrial Accidents

Azerbaijan is planning to accede to the *Convention on the Transboundary Effects of Industrial Accidents* by the end of 2003. The Convention was therefore translated and the approved Azerbaijani version was presented to the Ministry of Foreign Affairs.

A new law based on the EU Seveso II Directive and the Convention's provisions is currently being drafted. Furthermore, national procedures to comply with the Convention are already in place. Work in this area is still in its early stages, for instance no emergency information systems have been established yet. A national list of hazardous substances whose handling requires permission has been drawn up.

Convention on Access to Information, Public Participation in Decision-making and Access to Justice in Environmental Matters

Azerbaijan was one of the first to sign and ratify (in 2000) the UNECE Convention on Access to Information, Public Participation in Decision-making and Access to Justice in Environmental Matters (see Chapter 3 on information and public participation).

In 1998-2002, in the process of democratization, Parliament adopted laws setting out mechanisms to exercise many democratic rights and freedoms. This was envisaged in such laws as the 2000 Code of Civil Practice, the 2000 Code on Administrative Offences, the 1996 Handbook on Environmental Impact Assessment and the 1998 Law on Freedom of Information. This legislation also corresponds to the three pillars of the Convention.

Azerbaijan, with six other countries, participates in a project financed by TACIS to raise public awareness of environmental issues. OSCE is supporting the establishment of an information office in the Ministry to provide environmental information and guidelines to access it.

4.4 "Environment for Europe" process

Azerbaijan participated in the fifth Ministerial Conference "Environment for Europe" in Kiev in May 2003, and it was actively involved in developing the Environment Strategy for Eastern Europe, the Caucasus and Central Asia. This Strategy is intended to provide clear directions for environmentally sustainable policies at national and international levels and to serve as a basis for developing partnerships among all countries in the region.

Azerbaijan is also a member of the Working Group on Environmental Monitoring and, through the Working Group, contributed to its reports to the Kiev Ministerial Conference and to the Kiev Assessment Report prepared by the European Environment Agency.

4.5 Cooperation with the European Union

The partnership between the European Union and Azerbaijan came into effect in July 1999 with the conclusion of the Partnership and Cooperation Agreement.

Defining the priorities of TACIS in the country, the Indicative Programme 2004-2006 was discussed and approved by both Azerbaijan and the European Commission. Within its framework, €30 million will be granted for technical assistance, including a significant amount for tackling environmental problems. The focal organization for the EU TACIS technical assistance programme is the National Coordinating Unit for EU Technical Assistance in Azerbaijan under the National Coordinator. In the coming months, the work and discussions on identified projects will start.

Azerbaijan is currently also harmonizing its legislation with EU directives. The European Union, through the TACIS programme, is promoting environmental monitoring, standards and norms for the use of water resources. The Ministry of Ecology and Natural Resources is harmonizing Azerbaijan's water legislation with the EU Water Directive of 2000.

4.6 Other regional cooperation

Non-governmental and governmental networks

Azerbaijan is part of two networks in the South Caucasus: the Regional Environmental Centre for the Caucasus and the Caucasus Environmental NGO Network. They are ad hoc regional structures striving to broaden cooperation in tackling environment-related problems and ensuring sustainable development.

Caspian Environment Programme

Supported by UNEP, UNDP, the World Bank and the European Union, the Caspian Environment Programme was established by its five littoral States (Azerbaijan, Islamic Republic of Iran, Kazakhstan, Russian Federation and Turkmenistan) in 1998. It has established and put into operation a Project Coordination Unit and regional thematic centres, national Caspian action plans, a regional strategic action programme and a priority investment portfolio. The five littoral States also prepared and, in November 2003, adopted and signed the Framework Convention for the Protection of the Marine Environment of the Caspian Sea. The focus of the Programme is now on implementation of the SAP in three priority areas: biodiversity, invasive species and persistent toxic substances.

Other cooperation

Azerbaijan and Georgia have a memorandum of understanding (MoU) on collaboration in the development and implementation of the pilot project to monitor and assess the Kura river basin. Azerbaijan is preparing MoUs on environmental protection with Georgia, the Islamic Republic of Iran, Turkey and Ukraine. Azerbaijan is negotiating an MoU on the Samur river with the Russian Federation. The aim of these MoUs with neighbouring countries is to strengthen cooperation on the environment and nature conservation. Cooperation takes place primarily through exchanges of experience and expertise in science and technology, information and methodologies for public participation. Cooperation may also take the form of joint projects, such as wetland management with the Islamic Republic of Iran.

The North Atlantic Treaty Organisation (NATO), through its Science Programme, awards around 50 grants and fellowships to Azerbaijani scientists for collaboration on research or to study in NATO countries, such as grants for research into the Caspian Sea's circulation, especially as it affects potential oil spills, and for improvement of the quality of water supply, especially with regard to the Kura river. Azerbaijan has also started to set up working procedures with NATO for assistance in the clearance of unexploded ordnance (see chapter 12 on Human health and environment).

4.7   Bilateral and multilateral cooperation

Although some donors' activities have an environmental impact, most are related either to oil products or to humanitarian assistance, such as poverty reduction and support to internally displaced persons.

Azerbaijan has bilateral technical cooperation arrangements with several countries. On climate change prevention, the Canadian International Development Agency (CIDA) is training and providing capacity-building for the reduction of greenhouse gas emissions. France is involved in an afforestation project for carbon absorption. Finland and the EU are financing sustainable forest development and management. Germany participates in the conservation of the Ag-Gol reserve and the establishment of the Samur-Yalam national park. Japan, through a World Bank grant, is assisting institutional reforms in the Ministry of Ecology and Natural Resources. The United States Agency for International Development (USAID) finances the South Caucasus Water Management Project covering the Kura river.

4.8   Conclusions and recommendations

Azerbaijan is actively developing international environmental cooperation in many areas with other countries, international organizations and institutions. In addition to having signed and ratified a number of global and regional environmental conventions, Azerbaijan has established many bilateral and multilateral partnerships and has concluded numerous framework and sectoral agreements. In most areas, Azerbaijan is harmonizing its legislation with international and European norms, in accordance with the requirements of the international conventions that it has ratified as well as in view of its interest in joining the European Union. To further meet its international obligations, Azerbaijan has drawn up general and specific policy and action plans and sought foreign assistance in programme formulation and implementation. The principles of sustainable development are a good basis for integrating a large variety of related issues.

While the Ministry attaches importance to international legal instruments, implementing and complying with the new norms and action plans have not been a priority for all institutions concerned. An analysis will provide clearer goals and ultimately ensure a stronger commitment from the ministries involved. To improve the situation, strategic plans for implementation should be developed as soon as ratification is proposed. They should go beyond the mere translation of international commitments into national legislation and include funding commitments for implementation and compliance. Instructions, norms and standards as well as action plans should be used for the implementation of national laws and international agreements.

Recommendation 4.1:

(a) *The Cabinet of Ministers should give high priority to the implementation, compliance with and enforcement of international conventions and national laws by developing and putting in practice national environmental norms and standards, instructions and practical action plans following existing international commitments.*

(b) *The Ministry of Ecology and Natural Resources should assess the cost of implementation of a new international legal instrument for environmental protection before ratification in order to acquire the necessary resources.*

Although Azerbaijan ratified the 1989 *Basel Convention on the Control of Transboundary Movements of Hazardous Wastes and their Disposal* in 2001, the competent authority was only appointed in July 2003, resulting in delays in implementation. However, the Ministry of Ecology and Natural Resources is fully aware of its responsibilities and has already prepared some regulations and a "passport" for hazardous waste. Some further actions are forecast, such as an inventory and classification of hazardous waste.

Recommendation 4.2:

The Ministry of Ecology and Natural Resources should:

- *Speed up the development of a new law on the movement of hazardous waste based on the provisions of the Basel Convention;*

- *Set up the inventory of hazardous waste;*

- *Finalize the development of a classification system for hazardous waste based on the Basel Convention;*

- *Set up a database on the export, import and movement of hazardous waste in the country.and*

- *Develop a permitting system for hazardous waste.*

Priorities for improving transboundary waters include cooperation to reduce pollution by building treatment facilities, identifying more clearly the problems for different rivers and waters with the use of modern technology, improving assessment systems and databases to assist decision makers, and developing financial mechanisms to meet protection goals for different water bodies. The most urgent problem to be tackled is the protection of water from pollution by the two main rivers in the South Caucasus, the Kura and the Araz. In the framework of the memorandum of understanding with the Russian Federation on the Samur River, negotiations are ongoing.

Azerbaijan is working with Georgia on a different memorandum to improve the Kura River's conditions. Water quality in the Araz river causes serious concerns for Azerbaijan as a downstream country. In this regard, accession of all riparian States to international legal instruments in the environmental field is crucial for efforts on water quality improvement to be efficient.

Recommendation 4.3:

The State Committee for Amelioration and Water Management, in cooperation with the Ministry of Ecology and Natural Resources, and in consultation with the appropriate authorities of the other riparian countries, should take steps to establish an intergovernmental working group composed of high representatives of the riparian countries of the Kura and Araz rivers (Armenia, Azerbaijan, Georgia, Islamic Republic of Iran and Turkey) to cooperate on the sustainable management of these rivers. The intergovernmental working group should coordinate all projects, plans and development affecting water quality and quantity in the Kura and Araz rivers.

Figure 4.1: Map of Kura and Araz basins

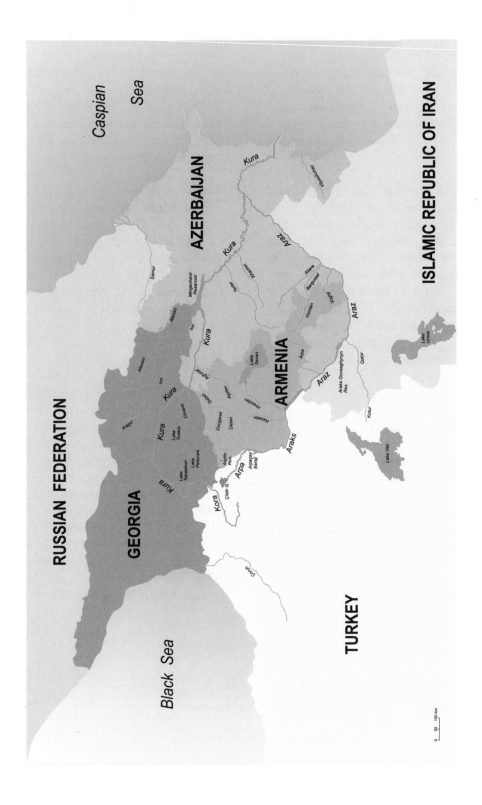

PART II: MANAGEMENT OF POLLUTION AND OF NATURAL RESOURCES

Chapter 5

AIR MANAGEMENTAND TRANSPORT

5.1 Recent developments and trends in air quality

Total emissions of air pollutants by stationary and mobile sources have fallen since 1990, due to the recession and reduced industrial activity. In 1990 total discharges into the air from stationary sources amounted to 2.1 million tons, while in 2002 this was only 217,000 tons. Air pollution is most severe in the capital city of Baku, an industrial hub of the former Soviet Union, as well as in other large cities such as Sumgayit, the chemical production centre. During the same period emissions from transport have increased following a rapid growth in car ownership and use. Traffic is burgeoning in cities, but vehicle registration, inspection and maintenance lag behind what is needed to support efforts to improve air quality. Poor fuel quality and ageing vehicles worsen the emission problems. In 2002 about 620,000 tons of pollutants were emitted into the atmosphere, of these, 65% were car exhaust gases. The situation will worsen if idle enterprises start working at full capacity again.

5.2 Trends in air emissions

In the past, large cities in Azerbaijan like Sumgayit suffered from air pollution levels considered unsafe for human health. The industrial collapse has improved air quality, but this has been partially offset by a rapid increase in transport. The main sectors contributing to air emissions today are transport, industry and energy. In 2002, 60% of total air emissions came from mobile sources, against 40% from stationary sources. Figure 5.1 shows the total emissions from mobile and stationary sources from 1990 to 2002. Compared with 1991, emissions from stationary sources in 2002 were 91% lower, and emissions from mobile sources 41% higher. Air emissions from stationary sources are given in table 5.1.

Azerbaijan's carbon dioxide (CO_2) emissions over the past decade have been declining, mostly because of the country's economic problems. CO_2 emissions from both petroleum and natural gas processing and use fell. Total CO_2 emissions from fossil fuel combustion decreased from 16.38 million tons in 1992 to 9.14 million tons in 2001.

The emission inventory system in Azerbaijan is based on annual emission reports that the operators of air polluting companies are obliged to provide. This reporting system already existed in the Soviet era. At present, due to the transition period in the country's economy, the forthcoming privatization and the break-up of larger units, the system suffers from some inconsistencies.

Although the inventories do not specify all the pollutants, it seems that in 2002 the main pollutants from stationary sources were hydrocarbons (124,000 tons), volatile organic compounds (5,300 tons), sulphur dioxide (13,600 tons), carbon oxide (18,200 tons), nitrogen oxides and dioxides (26,300), and particulate matter or dust (29,400 tons). The main stationary sources are oil refineries, electric power plants, chemical industries, construction industries, and the iron and steel industries.

Table 5.1: Air emissions of selected air pollutants from stationary sources

(in thousand tons)

	1990	1995	1996	1997	1998	1999	2000	2001	2002
Total	2,109	879	382	390	443	575	515	577	217
Particulates/dust	148	23	18	22	21	17	19	28	29
Liquid and gaseous substances	1,960	856	..	368	422	558	496	549	188
Of which:									
SO_2	90	50	40	38	35	37	35	15	14
NO_x	59	32	24	26	25	24	24	27	26
CO	71	22	19	23	21	22	26	28	18

Source: State Statistical Committee. Statistical Yearbook. 2002.

Figure 5.1: Air emissions from 1990 to 2002

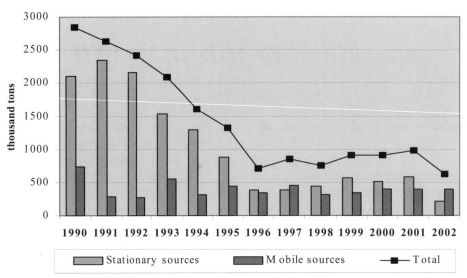

Source: State Statistical Committee. Statistical Yearbook. 2002.

Industry

Because of obsolete or non-existent technology and equipment for the purification of stack gases, the oil and gas industry is a major emitter of sulphur, carbon and nitrogen oxides. Among the biggest emitters of hydrocarbon are the Azerneftyanajag and Azerneftyag oil refineries. Most emissions originate from the flaring of fuel in the heaters. The refineries have been running well below capacity, with overall refinery utilization at 40%. Heating oil accounts for approximately 50% of the output of the refineries, followed by diesel fuel (28%), petrol (10%), motor oil (7%), kerosene (3%) and other products (2%). Both refineries are in need of modernization, which the Government estimates would cost between US$ 600 and 700 million. Recently, emissions were reduced through measures such as the installation of modern

equipment. Emission reductions of 3.5% (440 tons) were achieved in 2002 at the Azerneftyanajag and of 45% (2,364 tons) at Azerneftyag compared to the previous year (see Chapter 11 on environmental concerns in the oil and gas sectors for more details).

In Baku, emissions of industrial pollutants not related to the oil and gas industry are relatively low, except for the releases of dust from the Garadagh cement plant.

In Sumgayit, some of the pollutants include hydrogen fluorides from aluminium production, mercury and other toxic emissions from an amalgam plant for chlor-alkali production, heavy metals such as lead, zinc and cadmium in the dust from a steel plant, and various toxic substances from the petrochemical industries (producing polymers, propylene and ethylene).

Box 5.1:   Air pollution in Sumgayit

Sumgayit was founded in the 1950s as a centre for the chemical and petrochemical industries. It soon became one of the largest industrial centres of the former Soviet Union. Industrial areas occupied over 34% of the city, employing hundreds of thousands of workers. About 88 large facilities were built, of which 10 became heavy air polluters. Annual air emissions were about 100,000 tons. Emissions per square kilometre amounted to 1,200 tons in 1990-1991, while the average value for Azerbaijan was about 24 tons/km^2. Apart from the "classic" pollutants, toxic substances, mercury, chlorine, hydrogen fluoride and heavy metals were released into the ambient air, affecting the local population, especially sensitive groups. Persistent organic compounds, such as dioxins and dibenzofurans, were released from petrochemical industries. The city had one of the highest morbidity rates during the Soviet era. In 1992, Sumgayit was declared an environmental disaster zone, although air emissions had been declining since 1990. The city was later designated a free economic zone, in order to foster economic growth and the introduction of new technologies there. However, the problems of uncontrolled emissions, persistent pollutants and the liability for past pollution remain unsolved. In addition, although emissions have decreased, they continue to pollute the air.

Source: Caucasus Environmental Outlook, 2002.

Energy

Azerbaijan's power sector has an installed generating capacity of approximately 5.1 gigawatts (GW), consisting of eight thermal plants (accounting for 80% of the generating capacity) and six hydroelectric plants, all State-owned. Two thirds of the country's thermal capacity is powered by residual fuel oil, and by natural gas as the secondary fuel. Built during the Soviet era, Azerbaijan's power infrastructure is generally in poor condition, with minimal public investment and maintenance since independence. About half of the turbo-generators and boilers have been in use for more than 40 years and the results are high fuel consumption, low thermal efficiency and high emissions. Because of the country's inefficient distribution network, much of the generation is lost in transmission, making Azerbaijan a net electricity importer since 1994. In general, most thermal-electric power plants in Azerbaijan are relatively old and in need of modernization, and much of their equipment has deteriorated. The thermal power plants are largely fuelled by oil and emit mainly NO_x and SO_2.

Several projects are under way to restore and add new capacity to Azerbaijan's power sector. In 2002, a decree was adopted for the energy sector, setting the goal of eventually switching all thermal power plants to natural gas fuel. So far, a new 400 MW natural gas-fuelled expansion was completed for the Severnaya North State Regional Power Plant (also known as the Shimal Power Plant) and is now in operation. This reduces the emissions of NO_x and SO_2.

Mobile sources

Impact of transport on air quality

Urban air pollution is a matter of increasing concern in Azerbaijan, due to rapid urbanization,

motorization and economic growth. In many cities, transport is said to be the main source of air pollution. The total number of registered passenger cars was 271,000 in 1995, and increased by 22% to 331,000 cars in 2001 (see table 5.2). The ageing vehicle fleet and the lack of vehicle servicing, inspection and maintenance in combination with poor fuel quality all aggravate air pollution.

As a result of the downsizing of stationary sources and a drastic decrease in public transport, the share of mobile sources in total national emissions of NO_x, VOCs and CO has systematically increased. The total emissions from mobile sources amounted to 403 thousand tons in 2002, which is 65% of the total. The most common fuels are petrol and diesel. In addition, few cars have catalytic converters, yet this is by far the most effective means of reducing emissions of CO, hydrocarbons and NO_x from petrol-fuelled vehicles.

At the fourth "Environment for Europe" Ministerial Conference in 1998, Azerbaijan agreed to the regional strategy for phasing out added lead in petrol by 2005, and it has banned the use of leaded petrol since 1997. While a ban on the use and production of leaded petrol is the major step towards phasing out added lead in petrol, enforcement and control are also important in ensuring a complete ban. However, there are reports that lead and other additives are being added illegally to the petrol pool in the distribution system, indicating that the official statistics of the market share of unleaded petrol may overstate actual progress.

5.3 Trends in air quality

In general, the ambient air quality in Azerbaijan has improved in recent years as a consequence of the sharp decrease in air emissions. The main air polluting sectors, traffic and industry, are concentrated in the largest cities.

Table 5.2: Number of motor vehicles

(in thousands)

	1990	1995	1997	1998	1999	2000	2001
Total	398.8	392.2	374.7	392.8	409.3	440.6	451.6
Lorries	99.5	79.7	71.9	79.9	69.7	78.6	77.1
Buses	14.0	12.8	12.1	13.7	14.9	16.8	17.3
Cars	260.2	278.3	271.3	281.3	306.9	332.0	343.0
Motorcycles	..	13.3	3.2	9.3	9.3	6.4	6.7
Other	25.0	21.4	19.4	17.9	17.7	13.3	14.3

Source: State Statistical Committee. Statistical Yearbook. 2002.

Air quality data are collected by the National Department of Environmental Monitoring within the Ministry of Ecology and Natural Resources and by the sanitary and epidemiology centres of the Ministry of Health. The monitoring data are analysed and put in a database, and the information is distributed among other interested ministries and institutes. Furthermore, the data are published annually in the Statistical Yearbook.

Several air pollutants are monitored daily in eight cities (see table 5.3). The monitoring network has a relatively small number of stations with a limited mode of operation and the analyses of air quality do, therefore, not reflect Azerbaijan's complete air quality situation.

Only total suspended particulates (which include coarse particulates having no significant impact on human health) have been measured, and no data are available for fine particulates (PM_{10} and $PM_{2.5}$); thus, there are many shortcomings in the measurement of total suspended particulates. Heavy metals in the air are only occasionally measured.

Air quality monitoring is based on air quality standards, the so-called maximum allowable concentrations (MACs) inherited from the Soviet era. MACs are determined for different substances

and are in general more stringent than WHO or EU standards. Table 5.3 shows the national air quality standards for selected pollutants.

Monitoring data from 1991-1995 show that the levels of several pollutants in the cities of Mingechevir, Ali Bairamli, Ganja, Sumgayit and Baku were up to five times higher than the standards. According to the National Department of Environmental Monitoring, the concentrations of nitrogen oxides (NO_x), furfural, dust and suspended particulates exceeded the maximum allowable concentrations in 2002. The data for the first part of 2003 again show that concentrations of dust, NO_x, soot and furfural breach the MACs.

5.4   Monitoring of air quality

The legal provision for monitoring is in the Law on Environmental Protection and the Law on Air Protection. Three types of air monitoring take place: monitoring of industrial emissions, monitoring of background emissions and monitoring of air in inhabited areas. The Ministry of Ecology and Natural Resources, through its National Monitoring Department, is responsible for the first two, while the Ministry of Health and its sanitary and epidemiology centres monitor air quality in populated areas.

Table 5.3: Selected national air quality standards

Pollutant	Maximum allowable concentration (in mg/m^3)	
	For a given moment (maximum)	For 24 hours
Dust	0.1500	0.0300
Sulphur dioxide	0.3000	0.2000
Carbon monoxide	3.0000	2.0000
Nitrogen dioxide	0.0800	0.0700
Nitric oxide	0.4000	0.2400
Chlorine	0.0500	0.0300
Mercury	0.0006	0.0003
Sulphuric acid	0.3000	0.1000
Formaldehyde	0.0200	0.0050
Furfurol	0.0300	..
Hydrogen sulphide	0.0080	..
Hydrogen fluoride	1.0200	0.0030
Lead and its compounds (except tetraethyl lead)	0.0010	0.0002

Source: Ministry of Ecology and Natural Resources, 2003.

The National Department of Environmental Monitoring is the result of a merger of all separate research and monitoring centres. It measures air pollution three times a day, at 7 a.m., 1 p.m. and 6 p.m. There are in total 26 stations for air pollution in 8 cities: 8 stations in Baku, 3 in Sumgayit, 5 in Ganja, 4 in Mingechevir, 1 Sheky, 2 in Ali-Bairamli, 1 in Lenkeran, and 2 in Nakhchivan (see table 5.4).

The State Hydrometeorological Service measures meteorological parameters, such as temperature, deposition through rain and snow, cloudiness, sunny days and wind speed, at 60 stations throughout the country, 3 to 4 times a day. In addition the National Department of Environmental Monitoring also measures meteorological parameters at 26 stations.

The Ministry of Health and its 11 sanitary and epidemiological centres measure water and air quality. Their stations are located in cities, and measurements are taken once a week and additionally at the request of citizens. The pollutants measured by the centres are dust, suspended particulates (soot), hydrocarbons, carbon monoxide, nitric oxide, nitrogen dioxide, sulphur dioxide, hydrogen fluoride, chlorine, ammonia, furfural, formaldehyde, mercury and other pollutants (see table 5.4). Every

month the monitoring data are sent to the Ministry of Ecology and Natural Resources and the State Sanitary Epidemiological Service of the Ministry of Health.

5.5 Policy objectives and management practices

Policy framework

Environmental pollution continues to impose high economic and social costs on Azerbaijan. The Government approved the National Environmental Action Plan (NEAP) in January 1998. It sets the following objectives in air protection:

- Regional qualitative improvement of air in Azerbaijan;
- Inventory of the industrial sources of air emissions;
- Identification of ozone-depleting substances used and substances included in the Montreal Protocol and the Vienna Convention; and
- Improvement in air quality in populated areas.

The NEAP contains numerous technical, project-related recommendations, such as the purchase of equipment for laboratories, to control air pollution. However, it includes neither a clear timetable nor the budget required for each of these projects. Only few projects have been implemented.

Table 5.4: Air pollution monitoring stations in cities

City	Number of stations	Pollutants
Baku	8	Dust, sulphur dioxide, carbon monoxide, nitrogen dioxide, nitric oxide, hydrogen sulphide, suspended particulates, solvable sulphates, solid fluorides, hydrogen fluoride, chlorine, mercury, sulphuric acid, furfurol, formaldehyde, ammonia
Sumgayit	3	Dust, sulphur dioxide, carbon monoxide, nitrogen dioxide, nitric oxide, hydrogen fluoride, chlorine, mercury, sulphuric acid, furfurol, ammonia, hydrogen chloride
Ganja	5	Dust, sulphur dioxide, nitrogen dioxide, nitric oxide, hydrogen sulphide, solvable sulphates, solid fluorides, hydrogen fluoride, sulphuric acid
Mingechevir	4	Dust, sulphur dioxide, carbon monoxide, nitrogen dioxide, nitric oxide, solvable sulphates, phenol
Sheky	1	Dust, sulphur dioxide, carbon monoxide, nitrogen dioxide
Ali Bairamli	2	Dust, sulphur dioxide, nitrogen dioxide, carbon monoxide
Lenkoran	1	Dust, sulphur dioxide, carbon monoxide, nitrogen dioxide
Nakhchivan	2	Dust, sulphur dioxide, nitrogen dioxide, carbon monoxide

Source: Ministry of Ecology and Natural Resources, 2003.

Another important policy document containing objectives for air management and air quality is the National Environmental Health Action Plan (NEHAP) of 2001. It foresees the following actions:

- Rehabilitation and replacement of old gas and dust filters in industrial enterprises;
- Adjustment of the air quality standards;
- Establishment of environmental control stations equipped with gas analysers at town entrances to check vehicles' technical condition in terms of exhaust toxicity, and subsequent services to correct malfunctions;
- Monitoring of air quality in large towns with a common databank;
- Development of mathematical model to forecast air pollution, define toxic components in the air and map air pollution;
- Establishment of effective permanent public warning system and health institutions monitoring air quality in residential areas and other zones in towns where an increase in air pollution might damage public health;
- Research into acceptable limits for new chemical compounds, biological agents and environmental physical factors; and
- Informing the population about air quality in towns and industrial centres.

Recently, the National Programme on Environmentally Sustainable Socio-economic Development was adopted. It determines the main areas of sustainable development and includes a plan of action for 2003-2010. The National Programme includes measures to address air pollution, but has neither a clear timetable nor a budget for the different projects:

- Rehabilitation and replacement of gas and dust filters in industrial enterprises;
- Full switch to the use of unleaded petrol;
- Prohibition of old non-compliant transport means;
- Introduction of environmentally friendly transport, better pedestrian areas within cities;
- Improvement of air quality through expanding green areas; and
- Ensuring imported vehicles are equipped with catalytic converters and comply with EU norms.

Legal framework

The Law on Environmental Protection of 1999 provides the basis for developing and implementing programmes to combat air pollution. In 2001 a new Law on Air Protection was adopted. It sets out in seven chapters the requirements for monitoring, organization of activities, responsibilities of institutions, control and inspections, court procedures and international cooperation. It authorizes two types of norms: sanitary and ecological norms of air quality and impact on it, and technical norms of emissions into the air. For the latter, maximum permissible emissions levels are established. The Law replaces the old Law on the Protection of the Atmosphere of 1981. Furthermore, it foresees the issuance of 13 regulations with detailed procedures for air protection. The regulations have now been adopted (last one in April 2003), but are not yet implemented.

In addition, the Law on Hydrometeorological activity, adopted in April 1998, is the legal basis for hydrometeorological measurements and environmental monitoring.

The existing air quality standards were established according to the former Soviet GOST standards. Generally, these standards are based on maximum allowable concentrations (MACs), as in the former Soviet Union. According to the new Law on Air Protection, Azerbaijan will introduce new international standards. The switch from GOST standards to international (EU or WHO) standards will be complicated due to the old and deteriorated equipment.

Two types of standards are applied for fuel quality and pollutants in motor vehicle exhaust. The standards for emission levels from petrol engines are based on GOST 17.22.03 of 1987 on hydrocarbons and carbon monoxide content, and the standards for emission levels from diesel engines are based on GOST 21.393 of 1975 on smoke content.

Standards for stationary air pollution sources are also based on MACs. According to the present legislation, all enterprises whose activities generate air pollution need to receive permission from the Ministry of Ecology and Natural Resources. A so-called ecological passport is developed by an enterprise at its own expense and approved by the Ministry of Ecology and Natural Resources, where it is registered. The ecological passport is a normative technical document containing data on the company's use of resources and an assessment of the environmental impact of its manufacturing activities (see chapter 1 on policy, legal and institutional framework). The "maximum allowable emissions limit document" is set for every pollution source in an industry and varies depending on the type of industry, its location and the type of pollutant. When the maximum emissions limits cannot be achieved for some objective reasons, the industry is requested to decrease concentrations in stages until the maximum allowable emissions are reached.

Institutional framework

The Ministry of Ecology and Natural Resources is responsible for developing environmental policy, drafting legislation and regulations, implementing international agreements and conventions, and also for monitoring the state of the environment. Within the Ministry, several bodies have specific responsibilities relating to air:

- The Department of Environmental Policy and Environmental Protection;
- The National Department of Environmental Monitoring;
- The Hydrometeorological Service;
- The Climate Change and Ozone Centre, established in March 2003, within the Hydrometeorological Service. The Centre is working on the implementation of the United Nations Framework Convention on Climate Change, develops projects on climate change and coordinates the dialogue among experts working on climate change issues; and
- The State Control Inspectorate for Environment and Use of Natural Resources, which checks the compliance of industry with current legislation, including legislation on air emissions.

The 29 regional departments for environment and natural resources under the Ministry are the main control bodies. Their responsibility is to inspect and monitor the implementation of legislation on environmental protection. The regional departments are obliged to report their activities to the appropriate departments at the Ministry and to the regional office of the State Environmental Inspectorate at the Ministry on a monthly basis.

Other ministries and State institutions that have specific responsibilities with regard to air management include:

- The Sanitary Epidemiological Service of the Ministry of Health, which is responsible for the protection of the atmosphere in urban areas and the protection of the population against harmful effects;
- The State Statistical Committee, which reports annually on the condition of the atmosphere;
- The recently established Ministry of Transport;
- The Ministry of Internal Affairs through the State Traffic Police, together with the Ministry of Ecology and Natural Resources, which inspects the observance of legal requirements for environmental pollution from car exhaust; and
- The executive power of Baku, Department for Transport.

Economic instruments

The key principles for the introduction of economic instruments for environmental protection are laid down in the Law on Nature Protection and Environmental Management of 1992 and Cabinet of Ministers Resolution 122 on the Payments for Nature Use in accordance with the Application of Charges for Natural Resources, Discharge of Pollutants to the Natural Environment and Rates of Charges for Environmental Pollution. The new Law on Environmental Protection of 1999 reconfirms the legal validity and basis for economic instruments and its article 23 specifies the principles for using economic incentives for environmental protection by means of charges for natural resource use and environmental pollution. The only instrument aimed at controlling air pollution from stationary sources is the charge on air pollution.

The system of air pollution charges is very similar to the system used in the former Soviet Union. The charges were introduced in 1992 and are levied on 88 pollutants, according to their toxicity. The charge is paid every quarter, and the amount due is based on expected or actual emissions. The non-compliance fee applies when a company's emissions exceed the allowable limits, and the fee is 5 times higher than the base rates of the air pollution charges. The environmental effect of the charge and the non-compliance fee is rather limited due to their low tariff levels, the failure to enforce the legal provisions and inefficient collection.

In addition to these instruments, there are several transport-related taxes that affect transport as well as the environment. These are: taxes on transport fuels; taxes on the purchase, import and ownership of vehicles; and a road tax.

The Criminal Code of 2000 defines ecological crimes. Its article 251 defines air pollution crimes (see chapters 1 on policy, legal and institutional framework, and 2 on economic instruments, environmental expenditures and privatization for more details).

5.6 Policy objectives and management practices: Transport

The decision-making framework

In June 2003 the President signed a decree that is the basis for the establishment of the Ministry of Transport, which had been delayed for several years owing to internal disagreements and negotiations about its structure. On 28 August 2002, a Minister was appointed, but the Ministry's organizational structure and responsibilities are only now being approved. The new Ministry will consist of different former State bodies, such as the Transport Committee (*Azeravtonagliyyat*), the State Railway Company, and the Caspian Sea Shipping Company. Other State enterprises will be privatized.

At the moment there is no strategic planning document that outlines the goals and policies of the development of the transport sector. The Cabinet of Ministers has an "activities plan" on transport development, which is being transferred to the new Ministry of Transport.

Transport and environment policies are emerging from outside the Ministry of Transport, albeit in fragmented form. The National Environmental Action Plan (NEAP) and National Programme on Environmentally Sustainable Socio-economic Development include transport-related actions. The latter presents the following measures for the sustainable development of the transport sector:

- Adoption of regulations regarding the negative impact of vehicles on the environment;
- Improvement of legislation to ensure harmonized policy in the transport sector;
- Switch to environmentally cleaner fuels for vehicles;
- Reconstruction of highways and railways to meet international standards;
- Reduction in urban traffic density;
- Full Switch to the use of unleaded petrol in transport; and
- Prohibition of the use of old and heavily polluting vehicles.

The institutions responsible for the implementation of these measures are the Ministry of Ecology and Natural Resources and the Ministry of Transport. The NEAP of 1998 identifies air pollution from transport as a critical environmental issue, and its recommendations are in line with the actions in the National Programme on Environmentally Sustainable Socio-economic Development. However, it is unclear to what extent the recommendations are being implemented.

Owing to its favourable geographic location, Azerbaijan is developing into a major transit country between Central Asia and Europe as well as between the Russian Federation and the Islamic Republic of Iran. As a consequence of the lack of investments over the past decade, the transport infrastructure is in urgent need of rehabilitation and modernization. The European Union's Transport Corridor Europe Caucasus Asia (TRACECA) programme has been active in the Caucasus since 1993, upgrading and integrating rail and road transport networks in an effort to promote trade ties with Europe along a West-East corridor. In Azerbaijan, the TRACECA programme is providing logistical support to help the ageing rail system meet the country's growing oil transit needs.

Public transport in Baku

Urban transport in Baku faces the twin problems of dealing with the rapid growth in car traffic and maintaining a viable public transport system. The total emissions from mobile sources in Baku reached 285,000 tons in 2001, which is 71% of total emissions from mobile sources in the country.

Public transport in Baku relies on 300 buses, 8 to 10 trams and 2 metro lines. In addition, about 2,600 to 2,700 private minibuses are operating in the capital, and this number is still growing. The vehicles are very poorly maintained and do not undergo regular technical inspections. The Transport Department of Baku prepared an annual programme for the sustainable development of public transport, which contains such proposals as replacing old buses, reducing the number of minibuses, introducing EU vehicle emission standards (Euro 2 and Euro 3 standards), diverting heavy transport from the city, and introducing stricter controls of standards and technical requirements. To date, however, none of these measures has been implemented.

The Transport Department is negotiating with the Japan International Cooperation Agency (JICA) to start an urban transport improvement plan up to the year 2020, which will identify priority projects for transport development in Baku.

5.7 International agreements and activities

United Nations Framework Convention on Climate Change

Azerbaijan signed United Nations Framework Convention on Climate Change in 1992 and ratified it in 1995. In 2000, it ratified the Kyoto Protocol to the Convention. For the implementation of the Convention, the State Commission on Climate Change was established in 1997 by presidential decree. The Commission is chaired by the Deputy Prime Minister and includes the heads of governmental and private institutions.

Azerbaijan prepared its first National Communication to the United Nations Framework Convention on Climate Change and presented its emission inventories of greenhouse gases for the years 1990-1994. The data are in compliance with the Intergovernmental Panel on Climate Change's methodology and cover the emissions of three greenhouse gases – carbon dioxide (CO_2), methane (CH4) and nitrous oxides (N_2O) – as well as gases with an indirect greenhouse effect, such as nitrogen oxides (NO_x), carbon monoxide (CO), non-methane volatile organic compounds (NMVOCs) and sulphur dioxide (SO_2). Azerbaijan has not published or prepared complete emission inventories since 1994.

In 2000, with financial support from the Global Environment Fund (GEF), national experts implemented phase II of the project "Measures to Build Capacities in Priority Areas", which is a continuation of the first National Communication. The objectives were to determine the technical needs for capacity-building and to assist with capacity-building to participate in the climate monitoring network. Currently Azerbaijan is preparing to apply for GEF assistance to prepare its second National Communication.

Convention for the Protection of the Ozone Layer

In June 1996 Azerbaijan ratified the Vienna Convention for the Protection of the Ozone Layer and the Montreal Protocol on Substances that Deplete the Ozone Layer. The Ministry of Ecology and Natural Resources is the national coordinating body that develops and establishes the necessary regulatory and legal framework to control the trade in and use of ozone-depleting substances (ODS), to enable Azerbaijan to fulfil its obligations under the Montreal Protocol. The National Ozone Centre was established to help it phase out ODS. The initial country programme for the phase-out of ODS was compiled in 1997, and several projects in cooperation with GEF have been implemented since then.

The total consumption of ODS in Azerbaijan decreased from 966 metric tons of ozone-depleting potential (ODP) before ratification of the Protocol to 13.6 metric tons ODP in 2002, fulfilling the obligations under the Montreal Protocol. This drop of almost 99% has been achieved through structural changes in industry and a significant decrease in the production of refrigeration equipment. It is foreseen that ODS consumption will continue to fall and that by the year 2005 it will be negligible or nil.

The Chinar refrigerator plant in Baku and the Sumgayit compressor plant are phasing out ODS through initiatives aimed at recovery/recycling of refrigerants and establishment of halon banking.

Convention on Long-range Transboundary Air Pollution

Azerbaijan has been a Party to the UNECE Convention on Long-range Transboundary Air Pollution since 2002, but has not ratified any of the Protocols. The Government intends to ratify the Protocol on Persistent Organic Pollutants (POPs), the Protocol on Heavy Metals, and the Protocol on Long-term Financing of the Cooperative Programme for Monitoring and Evaluation of the Long-range Transmission of Air Pollutants in Europe (EMEP) in the near future.

5.8 Conclusions and recommendations

The Law on Environmental Protection of 1999 provides the basis for developing and implementing programmes to combat air pollution. In 2001 a new framework Law on Air Protection was adopted. It sets out the requirements for monitoring, organization of activities, responsibilities of institutions, control and inspections, court procedures, and international cooperation. The Law foresees the issuance of 13 regulations with detailed procedures for air protection. The regulations have been adopted (last one in April 2003), and implementation has begun. The implementation of all these regulations for the Law on Air Protection will complete the modernization of air protection legislation in Azerbaijan.

The new Law on Air Protection also calls for changing the ambient quality standards from the old GOST standards to those consistent with international guidelines and standards such as the health-based air quality guidelines of the World Health Organization (WHO). The conversion of GOST standards into internationally accepted standards would be complicated and would require both training and financing. The standards require not only changes in quantitative values, but also changes in the whole data collection, processing and analysis systems, which is resource- and time-consuming.

Recommendation 5.1:
(a) The Ministry of Ecology and Natural Resources should as soon as possible, undertake the necessary actions to implement the regulations for the Law on Air Protection, in order to enforce air protection legislation in Azerbaijan.
(b) Consistent with the new Law on Air Protection, the Ministry of Ecology and Natural Resources, together with the Ministry of Health, should adopt and implement new air quality standards and emission standards for stationary sources. The air quality standards should be in line with WHO air quality guidelines. The necessary training, equipment and financial resources should be made available to facilitate the transfer to these new standards.

Urban transport faces the twin problems of dealing with the rapid growth in car traffic and maintaining a viable public transport system. As a result of the downsizing of stationary sources and a drastic decrease in public transport, the share of mobile sources in total national emissions of NO_x, VOCs and CO has been systematically increasing. The total emissions from mobile sources amounted to 403,000 tons in 2002, which is 65% of the total. The ageing vehicles and their lack of servicing, inspection and maintenance in combination with poor fuel quality all aggravate air pollution.

Emerging traffic-related problems, such as pollution and congestion in major cities, deserve special attention. Appropriate measures to overcome them include the promotion of public transport (metro, trams, trolleybuses, regular buses and minibuses) combined with traffic management measures such as parking regulations and charges.

In addition, it will be important to establish an effective vehicle inspection and maintenance programme. The Ministry of Ecology and Natural Resources should play a coordinating role with the Ministry of Internal Affairs for vehicle inspections and emission testing. There are 35 vehicle inspection stations throughout the country, and only a few are equipped to measure technical vehicle requirements and fuel quality. However, purchasing more equipment will not in itself be effective unless steps are taken at the same time to identify high emitters and carry out corrective steps. The most urgent requirements for a good inspection and maintenance system are available service and repair facilities with good diagnostic equipment and qualified technicians and an up-to-date vehicle registration system.

Recommendation 5.2:
(a) The Ministry of Transport, in cooperation with the Ministry of Ecology and Natural Resources, should develop a sustainable transport strategy that fully incorporates environmental considerations. The strategy should address the traffic problems of air pollution and congestion in major cities with the appropriate measures.
(b) The Ministry of Ecology and Natural Resources together with the Ministry of Internal Affairs and its State Traffic Police should use resources from the State budget and other environmental funds to set up an effective vehicle inspection and maintenance programme in order to achieve emission reductions from the privately owned vehicle fleet. As part of this

programme, service and repair facilities with good diagnostic equipment and qualified technicians should be established.

At the moment, Azerbaijan uses the fuel quality and vehicle emissions standards from the former Soviet Union (GOST). These standards are mostly outdated and do not reflect what is now considered common international practice. The harmonization of fuel and vehicle emissions standards with internationally accepted standards should be considered. The old Soviet fuel and vehicle emissions standards can no longer meet the requirements of changing vehicle fleets and the imperative of protecting public health. Harmonizing these standards makes for greater efficiency in vehicle manufacturing and facilitates trade in refined products. The European Union adopted emission standards for petrol-fuelled cars in the early 1990s (so-called Euro standards) and has gradually tightened them. Similar requirements were adopted for diesel cars.

Recommendation 5.3:
(a) The Ministry of Transport, in cooperation with the Ministry of Ecology and Natural Resources, should develop, adopt and implement new emission standards for new mobile sources according to relevant European Union emission standards (Euro standards). In addition, adequate vehicle emission control schemes should be set up to check compliance with these standards.
(b) The Ministry of Fuel and Energy, in cooperation with the Ministry of Ecology and Natural Resources should adopt and implement, step-by-step, new fuel quality standards. Adequate fuel quality schemes should be set up to control the content of sulphur in diesel fuel and the content of lead in petrol fuel.

Reducing pollution and damage to human health cost-effectively requires an integrated approach to urban air quality management. An important step in developing an urban air quality management strategy is to be able to monitor and evaluate air quality. A good monitoring and modelling system is essential for policy-making suited to the primary objective of protecting human health. There are several key tasks for understanding the nature of urban air pollution, above all collecting data on ambient pollutant concentrations and developing an emissions inventory.

Most monitoring stations in Azerbaijan appear to be measuring CO, SO_2, NO_x and total suspended particles (i.e. particles of all sizes) regularly. However, no data are available for fine particulates (PM_{10} and $PM_{2.5}$), although they are far more damaging to public health than suspended particulates. In addition, ground-level ozone is not monitored in big cities where ozone levels are high. Ground-level ozone originates from transport emissions and could form summer smog. The six most important pollutants to monitor regularly are what the World Health Organization terms the "classical" pollutants: lead, $PM_{2.5}/PM_{10}$, carbon monoxide, sulphur dioxide, nitrogen dioxide and ozone.

It is also important to publish comprehensive annual air emission inventories, and this is not currently done. Inventories of greenhouse gases after 1994 are not yet available, although the Climate Change and Ozone Centre plans to prepare its second National Communication in the near future, including air emission inventories from 1994 onwards.

In addition, there is no quality control of emission inventories of air pollutants annually presented in the statistical yearbook, and inventories of mobile sources do not include a specification for different pollutants. All sources of emissions, mobile and stationary, should be identified in order to know the magnitude of the air pollution problem. A useful tool to improve the emissions inventory is the methodology of CORINAIR, which has been harmonized with that of the Cooperative Programme for Monitoring and Evaluation of the Long-range Transmission of Air Pollutants in Europe (EMEP).

Recommendation 5.4:
(a) The Ministry of Ecology and Natural Resources, together with the Ministry of Health should gradually establish a system of continuous monitoring of the six "classical" pollutants (lead, $PM_{2.5}/PM_{10}$, carbon monoxide, sulphur dioxide, nitrogen dioxide, ozone), to permit direct comparison with international guidelines and standards.
(b) The Ministry of Ecology and Natural Resources should start submitting complete air emission inventories as soon as possible, following the methodology of CORINAIR and the Cooperative Programme for Monitoring and Evaluation of the Long-range Transmission of Air Pollutants in Europe (EMEP).

Azerbaijan ratified the Convention on Long-range Transboundary Air Pollution in 2002, but did not ratify any of its Protocols. The main reason is said to be the lack of information on the national situation and also the lack of resources and equipment. Azerbaijan intends to ratify three Protocols (POPs, heavy metals, EMEP) in the near future. Participating in the international specialist work within the framework of the Convention could assist in training Azerbaijani air specialists. But before Azerbaijan ratifies the Protocols, it should carry out an analysis of the feasibility of ratification and develop appropriate plans and strategies for implementation of the Protocols.

Recommendation 5.5:

(a) *The Ministry of Ecology and Natural Resources should develop appropriate strategies for the ratification and implementation of the Protocols to the UNECE Convention on Long-range Transboundary Air Pollution.*

(b) *The Ministry of Ecology and Natural resources should raise its need to develop air quality monitoring and reporting to address requirements under the Convention with the Executive Body of the Convention, thereby seeking assistance from the Convention's programme centres and from the other Parties to the Convention*

Chapter 6

MANAGEMENT OF WASTE AND CONTAMINATED SITES

6.1 Introduction

The challenges before Azerbaijan for good waste management are many, and the Government has recently made waste management one of its priorities. The system of municipal waste collection, transportation and disposal works well in Baku city, but, in general, existing landfills do not meet international sanitary norms and standards, and there is insufficient waste separation. Rural areas are only partly covered by municipal waste service.

Outdated technologies remain sources of environmental pollution, including through the generation of solid industrial waste and sludge, but the quantity of generated industrial waste has decreased because, at present, the main industrial facilities, except those in oil and gas industries, work only at 5-30 % of their capacities. The main problem for the industrial sector is accumulated waste, disposed of at industrial sites or along with municipal waste at uncontrolled landfills.

The facility for storing obsolete pesticides as well as that for radioactive waste disposal are in poor condition and require rehabilitation. Medical waste needs to be separated and properly treated and incinerated. And action needs to be taken to clean up sites contaminated by exploration and exploitation of oil and gas along the coast of the Caspian Sea and nearby land, especially in the Absheron peninsula.

These are all formidable challenges, and the Government of Azerbaijan has been making great efforts to meet them, through new policies and strategies, new institutions and a range of actions aimed at good waste management.

6.2 Municipal waste

There is at present insufficient information within the country on the disposal of municipal waste. The Ministry of Ecology and Natural Resources intends to draw up an inventory in four cities (Baku, Sumgayit, Mingechevir and Ganja) of all waste, both newly generated and accumulated, but work on this has not yet started. Such information is essential in order to organize the treatment and environmentally sound disposal of waste and, especially, hazardous waste. Information is available on the quantity of municipal waste generated (see tables 6.1 and 6.2).

More information is available from Baku, where waste collection is reasonably good. For example, no municipal waste is disposed of at illegal dumpsites. The volume of municipal waste generated in Baku depends significantly on the season. During the summer, it is 48% higher than during the winter. The average volume of municipal waste is 54,822 m^3 in February against 144,842 m^3 in July. During summer all green waste (grass and bush cutting) as well as food waste from watermelons and melons are disposed of with municipal waste. In addition, because of the higher temperature, more drinks in plastic bottles are consumed and these are not recyclable. Also during summer tourism contributes to municipal waste generation.

In Baku, municipal waste contains flammable (80.8%) and non-flammable components (19.2%). Construction waste accounts for 14% of all municipal waste, and it is higher during summer. Industrial non-hazardous or treated waste, as well as waste from small enterprises, is also disposed of along with municipal waste in all cities.

As table 6.1 shows, the quantity of municipal waste generated in Baku alone was 641,435 tons in 2002. According to available data, the quantities of waste generated in other municipalities are very small (see table 6.2). This means that the reporting system in other cities and rural areas is not reliable and that many illegal dumps exist that are not taken into account. Setting up a good reporting system for municipal waste generation is the first step in improving its management.

Table 6.1: Quantity of municipal waste generated in main cities in 2002

City	Population	Total municipal waste generated	
		in tons	tons per person
Baku	1,818,000	641,435	0.353
Ganja	301,000	120,450	0.400
Sumgayit	288,000	166,667	0.579
Mingechevir	95,000	20,888	0.220
Ali-Bairamly	71,000	10,791	0.152

Source : Ministry of Ecology and Natural Resources, 2003.

Note : data in m^3 were converted into tons using the following ratio: 3 m^3 = 1 ton.

Three private companies, UP Azerbaijan, Kasko RCP and Kasko Waste Services, under contract with the communal department of the executive power of Baku are responsible for the collection, transport and disposal of municipal waste. The largest is UP Azerbaijan, which is a joint venture with a private German company. About 70% of its assets are in the hands of this German company (UP Gmbh) and 30% belong to the executive power of Baku. The company provides all necessary machinery and equipment for municipal waste transport and disposal. The company is hoping to make some profit in the future. Special containers are used for the collection of municipal waste and special waste trucks with compacting devices are used to take the waste to a landfill.

There is no separation of municipal waste, except for glass bottles and bread. Glass bottles are returned for reuse or put in separate containers. Bread is also put in special containers. Much of the municipal waste is in fact secondary raw material (see table 6.3).

In cities and towns outside Baku, municipal waste is collected and transported to landfills without any separation or treatment. This includes, for example, plastics, paper and metals that are potentially recyclable. There are no recycling facilities in Azerbaijan.

Baku relies on four landfills (see table 6.4). The Balakhany disposal site is the main facility for municipal waste disposal. At present, 7 out of an available 27 ha are used. Waste is pressed in the trucks and then compacted by bulldozers at the disposal site. The site is located about 25 km from the city and is closed off. About 110,000-120,000 m^3 of municipal waste is dumped there every year. The other three landfills are not used as intensively.

Each city and rayon has its own dump, but none of these is monitored or controlled, and there are no realistic projects for the construction of sanitary landfills. There are plans to construct one regional landfill, which would meet international standards, in Baku, but the practical work has not started.

Table 6.2: Quantity of municipal waste generated in some regions in 2002 in tons

Region	Population	Total municipal waste generated	Tons per person
Saliian	115,000	24,800	0.216
Imishli	107,000	4,500	0.042
Zagatala	109,000	3,650	0.033
Astara	87,000	487	0.006
Masally	179,000	456	0.003
Lenkeran	194,000	760	0.004
Sabirabat	141,000	7,961	0.056
Guichai	103,000	445	0.004
Neftchala	74,000	1,480	0.020
Khachmaz	148,000	16,873	0.114
Shamkir	177,000	1,000	0.006

Sources : Ministry of Ecology and Natural Resources, 2003.

Table 6.3: Resources of secondary raw materials and waste

(thousand tons)

	1990	1995	1999	2000	2001
Paper and paperboard	17.3	2.4	1.9	1.9	0.5
Secondary textile materials	14.9	6	1.3	0.4	0.3
Glass	37.4	1.2	10.7	0.1	0.7
Construction waste	35.6	34.6	20.1	23.7	25.8
Wood, thsd m^3	130.2	5.5	0.1	3.4	3.3
Secondary polymer	9.6	0.3	0.4	0.4	0.3

Source: State Statistical Committee. Environment. Statistical Yearbook. 2002.

There are no landfills that meet international sanitary standards and norms. Landfills are generally unorganized and unmonitored. They are neither fenced in nor guarded. After disposal, waste is not compacted nor covered with a layer of soil. At the dumps, waste is burned in the open, rather than in industrial incinerators, which leads to a diffusion of smoke in and around landfills. There is no protective layer at the bottom of the dumps to prevent hazardous substances from leaching into groundwater. Similarly, there are no measures for the collection and treatment of surface water that may also pollute groundwater. At the same time, there has been no research and analysis of the physical characteristics of the soil around the landfills.

The adverse effects of municipal landfills are:
- Heavy metals and toxic organic chemicals, which are formed from decomposing waste, penetrate the groundwater;
- Rain water also leaches soluble hazardous chemicals and penetrates into groundwater, which is used as drinking water;
- Contamination of air in the vicinity of landfills with volatile hazardous substances as well as smoke from uncontrolled waste-burning;
- As people and animals have unrestricted access to landfills, this creates various health risks, including potential meat contamination if those animals are used for food processing.

6.3 Hospital waste

Hospital or medical waste is disposed of together with municipal waste, after disinfecting contaminated material. In many hospitals, syringes and sharp materials are disposed of separately. (For more information on hospital waste, see chapter 12, on human health and environment.)

6.4 Industrial and hazardous waste, including obsolete chemicals

Pesticides

In recent years, the use of pesticides has almost stopped. According to the latest available statistics, only about 500 tons of pesticides were used in 2000, against more than 38,000 tons in 1988 (see table 6.5).

Notwithstanding the drastic decrease in the use of pesticides, the soil in many places is still contaminated by them. In the early 1990s about 30-40 kg of pesticides per hectare were used for cotton and about 150-180 kg per hectare for grapes. In Azerbaijan, the concentration of pesticides in the soil is higher that in other countries of the former Soviet Union. Azerbaijan does not produce pesticides at present, but imports small amounts of approximately 500 tons a year.

Table 6.4: Disposal facilities for municipal waste in Baku

Landfills	Year started	Total surface area in ha	Surface in operation,ha	Percentage of area in operation	Quantities transported per month, m^3
Balakhany	1963	200.0	27.0	13.5	111,390
Azizbekovo	1980	5.0	1.8	35.0	40,410
Surakhanski	1994	2.5	0.4	15.0	3,043
Karadagski	1994	25.0	3.0	12.0	5,516

Sources: Ministry of Ecology and Natural Resources, 2003.

Table 6.5: Total use of pesticides

(tons per year)

Year	1988	1990	1992	1995	1997	2000
Pesticides	38,671	33,889	33,540	5,375	938	500

Source: Hazardous Waste Management Agency, 2003.

The main problem is the stock of obsolete pesticides. In 1998, the organization responsible, Azerbaijan Selkhozkhimiya, was abolished and about 80 storage facilities remained without any supervision. At present, obsolete pesticides are stored in conditions that present a direct risk to public health and the environment. There are no special facilities for their storage. There is one site for the disposal of hazardous chemicals in Gobustansky region about 53 km from Baku. It is filled to capacity and is in very bad condition.

About 8,000 tons of pesticides, including DDT, are stored there in concrete containers, some of which have cracks or are open. This has resulted in the release of volatile compounds into the atmosphere. In addition, as rain falls, it washes the chemicals into groundwater. In some cases pesticides are stored outside the concrete containers without any precautionary measures. The population in the area has not been informed of the existing and potential hazards associated with this facility. No environmental impact assessment of the facility has been conducted, and there is no monitoring system. The site presents a particular health threat to the population of a small village about 5 km away from the facility. Another concern is a busy highway passing near the site that connects Baku and the town of Shemakha.

Analysis of food shows that the concentration of pesticides in some samples is higher than the maximum limits set in existing standards. The reliability of these data is questionable because of the use of outdated equipment and test methods. The number of food samples was 7381 in 1988 and 2200 in 1998. The problems with both pesticides and fertilizers were compounded after the privatization of land. Where before there had been strict State control over the use and distribution of agrochemicals, no new system was introduced to support private farmers with information and technical knowledge.

Hazardous waste

In general, waste and particularly hazardous waste has not been properly managed, although Azerbaijan is beginning to put more focus on this issue. A new draft national hazardous waste management strategy has been proposed (see below), and should be adopted in the near future. Adopting and implementing this strategy should improve overall hazardous waste management, including the institutional arrangements.

Statistical data on the generation of hazardous waste are not always reliable, and sources often give different figures. Table 6.6 is based on data received from the Ministry of Ecology and Natural Resources.

Table 6.7 gives the quantities of hazardous waste accumulated by 2002 in tons.

Table 6.6: Generation of hazardous waste in 1999

(tons per year)

Waste containing sulphur and sulphuric acids components	8,880
Waste containing mercury and its compounds	2,000
Oil sludges	1,100
Waste from stone cutting	413
Acid oil tar (gudrons)	389
Metallurgical sludges and slags	134
Waste containing vanadium	10
Asbestos waste	52
Others	52

Sources: Ministry of Ecology and Natural Resources, 2003.

Table 6.7: Waste accumulated by 2002 at some industrial sites

Oil processing plant in Baku	Waste containing kaolin and oil sludges – 4 million tons
Aluminium plant in Giadze	Red mud – 50,000 tons
Iodine plant in Baku	10,574 tons of activated radioactive coal
Pesticides in Gobustansky region	8,000 tons of pesticides

Source: Ministry of Ecology and Natural Resources, 2003.

Until now chemicals used in wine production, waste from machine tool manufacturing and obsolete chemicals have been dumped at unmanaged landfills. As a result, soil, air and groundwater are contaminated. Work on the environmentally sound management of hazardous waste has only just started. The system of hazardous waste management, especially its infrastructure, which covers testing procedures for hazardous chemicals, laboratory practices, the use of standard analytical methods for defining waste composition as well as technical guidelines on waste handling, still requires further development and improvement.

There is no organized system for the collection and disposal of hazardous waste. Enterprises transport and dispose of this waste themselves or they store it at industrial sites. In some cases hazardous waste is transported in open trucks, exposing both their drivers and the environment to contamination.

An inventory of all sources of hazardous waste was scheduled to be drawn up by mid-2002, but it has not yet been completed. Some enterprises report data on the amount of hazardous waste that they generate, but, in the absence of any legal requirement, this is not done regularly by all enterprises. All information concerning the generation of hazardous waste is provided annually to the State Statistical Committee, which processes and publishes the data in its statistical yearbook, as well as in a special publication, entitled "Environment".

One of the important tools for the identification of all sources of industrial waste, including hazardous waste, is an environmental audit of functioning enterprises as well as abandoned industrial sites. It provides realistic information to be used for developing concrete projects for industrial non-hazardous and hazardous waste recycling, reuse and disposal.

The Hazardous Waste Management Agency, which is under the Ministry of Ecology and Natural Resources, is responsible for the management of hazardous waste and has developed a "passport" for it. The "passport" is a one-page form, containing the following information: classification as per the Basel Convention, toxicity, general description and composition, chemical and physical characteristics, proposed treatment and use, and the names of the company and person responsible. The passport does not provide information on the possible result of any interaction with other types of waste or with the environment. Each type of waste is supposed to be accompanied by a separate "passport". The purpose of this "passport" is to serve as a reporting form for qualitative assessment. All companies that generate hazardous waste are responsible for filling in and submitting the "passport" to the Hazardous Waste Management Agency.

The existing classification system for waste dates back to the former Soviet Union. The system is outdated and does not clearly distinguish between different kinds of waste, especially hazardous waste. The draft national hazardous waste management strategy, if approved, would introduce a new classification system based on the provisions of the Basel Convention. It would cover environmental properties of hazardous waste and methods for their treatment. It would also hold companies that generate hazardous waste responsible for drawing up a waste reduction policy.

Oil drilling sludge

The landfill for oil drilling sludge is situated in Akhtarma, about 37 km from Baku. There are six long trenches in the form of a tunnel. The trenches are 170 to 205 m long, 4.5 m deep, 15.5 m wide at the top and 6.5 m wide at the bottom. The sludge contains 35% water, about 60% clay, 8% oil and a mixture of hazardous chemicals, which are used as agents in the oil drilling process. The site has a capacity of 45,000 m^3; at present about 27% is used. Sludge is transported by ships in special containers (volume 3 m^3) and then by large trucks to the disposal site. The storage facilities are insulated by a layer of concrete at the bottom to protect groundwater from contamination. The landfill is fenced and guarded, which minimizes access by animals and unauthorized persons.

However, there is a serious problem with run-off. The sludge is often stored in the open for some time before it is covered with a layer of concrete. Rainwater mixes with the sludge, and leaches hazardous chemicals and oil. The oil should be separated, but there are currently no resources available for this. There is also no monitoring system for the quality of groundwater, so it is impossible to know the extent of the contamination.

The oil resources are owned by Azerbaijan and their exploration and exploitation are under the responsibility of the State Oil Company of Azerbaijan (SOCAR). SOCAR has, in turn, signed over 20 major field agreements with approximately 30 companies from 15 countries. Among other things, these agreements require that oil exploration and drilling should be conducted according to international environmental standards. The foreign companies also carry out research to process sludge by sedimentation of heavy metals, separation of water-oil mixture and soil purification.

Little hazardous waste is recycled or reused. According to historical data, hazardous waste used to be recycled, but now virtually all of it is stored (see table 6.8).

In 2002 only 700 tons of hazardous waste was recycled or used. One of the few positive examples is the extraction of oil from sludge and oil-water mixture by some private companies. They treat hazardous waste chemically, separate the oil, decontaminate soil by mechanical and chemical methods, and reuse tyres and batteries. In the metallurgical sector, steel scrap is reprocessed.

Currently, hazardous waste is treated only in the oil sector. No other industry treats its hazardous waste. The main type of treatment in the oil sector is chemical treatment for waste neutralization by lime and solidification. (see table 6.9).

6.5   Radioactive waste

The main sources of potential exposure to ionizing radiation are medical diagnoses (X-rays), radioactive materials used by medical and research institutions, radioactive waste water in the oil and chemical industry and as well as radioactive devices in the military sector.

The geological nature of materials used in construction could lead at least in theory to indoor exposure to natural radon gas, but there is no equipment to assess this potential source of exposure. Another possible concern is radioactive contamination in residential areas near oil fields, where surveys have identified areas contaminated by radioactive (low specific activity (LSA)) emissions from the decay of natural radioactive isotopes present in rocks and gases, which concentrate in scales brought to surface when water is pumped into the pipes to maintain a steady oil flow. Experts of the Radiation Medicine Department suspect that the exposure of children living in these contaminated areas to radiation may lead to an abnormally high number of cases of leukaemia or other diseases, but no studies have been carried out so far to assess this potential problem. Additional concerns have been expressed in relation to the potential fallout from accidents that might occur in neighbouring countries where nuclear power plants are operated (Armenia, Islamic Republic of Iran, Russian Federation, Kazakhstan).

Table 6.8: Generation, treatment and disposal of hazardous waste

(thousand tons)

	1995	1996	1997	1998	1999	2000	2001	2002
Generation of toxic waste	27.0	18.0	36.1	33.9	13.0	26.6	16.4	9.8
Used or recycled toxic waste	25.4	5.3	8.1	30.5	14.8	11.4	0.4	0.7

Source: State Statistical Committee. Environment. Statistical Yearbook. 2002.

Table 6.9: Treatment of toxic waste in Baku and Sumgayit

(thousand tons)

	Treatment of toxic waste						
	1995	1996	1997	1998	1999	2000	2001
Baku	25.4	3.2	7.4	15.7	5.5	10.4	0.3
Sumgayit	..	1.2	0.7	14.7	9.3	1.1	0.1

Source: State Statistical Committee. Environment. Statistical Yearbook. 2002.

Temporarily used radioactive material can be imported with permission from the State Control Inspectorate for Environment and Use of Natural Resources (SCI) and the Prime Minister. At the moment, 482 facilities are mapped as sources of radioactive waste, and information on the type and quantity of isotopes used is available to the SCI and the Radiation Medicine Department of the State Sanitary Epidemiological Service. This Department is also involved in testing radioactive contamination of construction material and food products (although on a selective basis rather than on the basis of a regular monitoring programme), and in the control of occupational exposure. It also serves as the focal point for the international alert system of WHO.

International collaboration with the International Atomic Energy Agency (IAEA) on training and technical assistance (e.g. in the provision of equipment such as computers, spectrometers, dosimeters) started in 2000.

Since April 2003, Azerbaijan has participated in a programme (RAIS) to establish a computerized registry of radioactive material.

Radioactive waste is disposed of at a special facility (IZOTOP). This storage facility is located at approximately 37 km from Baku, in a relatively isolated area where there is no groundwater (up to a depth of 600 m) and 10 km away from the nearest water and gas pipelines. The facility was established in 1963, and since then has served as the only controlled landfill for radioactive waste in Azerbaijan. A grant worth €4 million received through the EU TACIS programme has helped to fund a major upgrade of the facility, which is expected to become operative by the end of 2004, bringing it in line with international good practices. After being sorted by activity and half-life, radioactive waste will be stocked in 200-litre barrels, which will be stored in 4-metre deep concrete bunkers and continuously monitored. The old dumping site will be liquidated under a "sarcophagus" of cement and asphalt. The TACIS grant also paid for a mobile laboratory for the detection and

analysis of radioactivity. The long-term sustainability of the operation and maintenance of the plan should be ensured through fees from private enterprises using this service. Public entities are not expected to be charged.

6.6   Contaminated sites

In connection with the inventory, discussed above, the Ministry of Ecology and Natural Resources intends to develop criteria for the rehabilitation of contaminated sites and to implement decontamination projects. The inventory is tentatively scheduled to be complete for 2004. Land polluted by oil in the Absheron Peninsula and at and around industrial sites, especially in Sumgayit, presents one of the most serious problems in terms of its adverse environmental and health effects.

Sumgayit industrial complex

Sumgayit has developed a 5-year ecological programme aimed at the environmental rehabilitation and restoration of the city and the closure of industrial sites. At present the main problem in Sumgayit is 200,000 tons of mercury sludge from chlor-alkali production that contains 0.2-0.3% mercury. It is stored at the industrial site in a special concrete reservoir. After sedimentation, it is removed and stored on land near the reservoir. Sludge is hazardous to the environment and to human health: first of all through mercury contamination of groundwater and soil, and secondly through the release of mercury vapour into the air. The health threats to the population of the city are particularly high because the industrial facility is situated about 5 km from the city (see chapter 12, on human health and the environment).

A treatment and storage facility for mercury sludge and hazardous waste is currently under construction near Sumgayit. The total area of this site is 55 ha, including 15 ha for mercury sludge. The facility, which is financed by the World Bank, will meet international environmental norms and standards.

An overall project to rehabilitate contaminated land and extract mercury from the earth is under development. The total cost of the project is US$ 8.1 million. It includes the construction of a treatment and storage facility for mercury sludge and other hazardous waste as well as the decontamination of soil polluted by mercury in and around Sumgayit's industrial sites. This project is now being implemented. A project to draw up an inventory of all industrial waste in the Sumgayit region is also under way.

Several other projects related to industrial waste management in Sumgayit are under consideration such as: environmental impact assessments for all new industrial facilities in the region, including the development and application of ecological passports, and ensuring that environmental norms and standards are applied, introducing a technology for chlor-alkali production that does not involve mercury-containing substances, and introducing a non-waste-generating technology for the production of polyethylene pipes.

Oil contaminated sites

Use of outdated technologies in oil production for many decades resulted in the contamination of soil with oil and mineralized water. At present, the contaminated area in and around the Absheron Peninsula covers approximately 10,000 ha, of which 7,400 ha is agricultural land. It has been confirmed that oil has soaked through the topsoil to a depth of up to 3 m and has leaked into the groundwater.

In many cases the same land is also contaminated by radionuclides. Gamma radiation in some of those areas is 20-50 times higher than the current maximum allowable limit. The land at refineries and petrochemical plants is also contaminated by oil and liquid hydrocarbons. All these contaminated areas are sources of air pollution, resulting from the release of volatile hydrocarbon compounds, which is particularly high in summer. The rehabilitation of this land is prohibitively expensive in the present economic situation.

At present there is very little oil production onshore. However, according to SOCAR, a programme has been developed for the rehabilitation in two stages of the land that was used for oil exploitation. In the first stage, the areas where oil is no longer produced (total

area circa 2,800 ha) will be cleaned up through the mechanical and biological treatment of the soil, after which the rehabilitated land may be used for crop production and pastures. The second stage includes only technical treatment for contaminated sites. This type of treatment will be used for land intended for housing or industry. Mechanical treatment includes drainage of artificial lakes and bogs, flattening the land surface and disposal of municipal waste at those sites. At present about 500 ha have been rehabilitated by mechanical treatment. The World Bank is supporting a pilot project for land rehabilitation by biological methods at the Tagieva oil-gas industrial complex.

Further application of the results of research and development in the form of concrete projects is needed to rehabilitate oil-contaminated territories, including their biological treatment.

Cleaner production

At present there are no programmes or activities concerning cleaner production technologies. New enterprises have to use only advanced technologies, which are also environmentally friendly. It should also be taken into account that developing and implementing a national cleaner production programme is one of the important tools to reduce the generation of industrial waste. Cleaner production technology keeps waste generation down. This approach is used at all new industrial facilities to solve the problem of industrial waste disposal, especially of hazardous waste. This particularly concerns the chemical, pharmaceutical and petrochemical industries, which are the main generators of hazardous waste.

In addition, cleaner production processes make it possible to manufacture high-quality products, which can compete internationally.

6.7 Policy objectives and management

Policy framework

In Azerbaijan's National Programme on Environmentally Sustainable Socio-economic Development (2003), prepared by the Ministry of Ecology and Natural Resources, the action programme addresses, inter alia, the urgent need to improve industrial and municipal waste management, including through the construction of landfills that meet health and environmental standards (see chapter 1, policy, legal and institutional framework).

The following measures are planned:
- Creation of a new landfill for hazardous industrial waste;
- Use of advanced practices for sorting, disposal and recycling of household waste;
- Introduction of appropriate methods for the generation of biogas from municipal waste;
- Construction of waste incinerators for the generation of energy and compost from waste.

Azerbaijan has recently completed a draft national hazardous waste management strategy. It is currently being reviewed and is scheduled for adoption by the Government by the end of 2003. Its main purpose is to provide a policy and decision-making framework for the environmentally sound management of hazardous waste. The strategy would establish a system for the collection, treatment, recycling and disposal of hazardous waste according to international practice and standards.

The strategy addresses both legal and institutional issues, human resources and capacities for implementation, public awareness, and waste reduction and recycling. It also includes key international principles for sustainable development and hazardous waste management. Should the strategy be adopted, special attention would be paid to the following issues:
- Waste minimization;
- Reducing environmental hazards and risks in the process of handling hazardous waste;
- Increasing reuse and recycling of hazardous waste;
- Industrial performance of major generators of waste, including environmental performance of industrial facilities: quality of products, effects on the environment and profit;
- Building new facilities for the environmentally sound disposal and storage of various types of waste, including medical waste.

Legal framework

The Law on Industrial and Municipal Waste (30 June 1998) addresses both industrial and municipal waste, including harmful gases, waste water and radioactive waste. Most of the focus, however, is on industrial waste. The only

provision concerning municipal waste is contained in article 12, which spells out all the necessary conditions for the construction of landfills (geological characteristics, distance from residential area) as well as sanitary and hygiene measures and norms to protect the environment. Unfortunately, in practice these measures are not implemented. The Law has not yet been supplemented by regulations, norms or standards. Developing and enforcing a comprehensive law on waste management is one of the main conditions for improvement in this area.

The draft national hazardous waste management strategy foresees the introduction of new legislation and regulations that would address a range of important issues, including the definition and classification of waste, an inventory of contaminated sites and their rehabilitation, a reporting system, licensing, and accident preparedness. Fines for non-compliance would be increased to ensure the implementation of the legislation.

Institutional framework

The executive powers of Baku and other cities are directly responsible for the collection, transport and disposal of municipal waste.

Enterprises that generate hazardous waste are responsible for its treatment, transport and disposal. The Ministry of Ecology and Natural Resources supervises municipal and industrial waste management. It issues permits to enterprises for the disposal of hazardous waste at specific sites. The Ministry also conducts inspections and has overall control functions for the sanitary and hygiene conditions of municipal and industrial waste sites, including for the general oversight of hazardous waste management.

The collection, treatment and disposal of medical waste fall under the responsibility of Ministry of Health.

In December 2002, the Hazardous Waste Management Agency was established under the Ministry of Ecology and Natural Resources. The Ministry is the competent authority for hazardous waste management, including the development and implementation of policies. Their staff requires training to be able to perform these functions.

Economic instruments

Most of the projects and activities for hazardous waste management are expected to be financed by long-term loans and grants from international donors. Once industrial facilities for hazardous waste treatment and disposal are installed, the enterprises that generate hazardous waste will be expected to pay annual fees and charges for waste treatment to the State Hazardous Waste Management Company, depending on the quantity and composition of the waste that they generate. The fees (and charges) will be increased to cover all operating costs. The cost of treatment and disposal of hazardous waste for which a specific source cannot be identified will be paid by the Oil Fund (see chapter 2 on economic instruments, environmental expenditures and privatization).

In 1992, the Council of Ministers adopted Resolution 122 (amended by Resolution 216 in 1993), which establishes the fines for discharging waste into the environment. The amount depends on the class of toxicity of the waste. If the waste exceeds the amount specified in the discharge permit, the excess waste is subject to a fine ten times the regular payment. The rates are:

	Manats per ton
• 1st class toxicity -	1500
• 2nd class toxicity -	1000
• 3rd class toxicity -	500
• 4th class toxicity -	250
• Non-toxic -	50

If the waste is taken to a controlled landfill, where there is no pollution of groundwater, soil and air, the payment is waived.

Payments for the disposal of municipal waste are regulated by order 1039 of the executive power (1997). Industrial enterprises and institutions pay 15,000-25,000 manats per m^3 of waste, while hotels and shops pay 30-60 manats per m^3 for the collection and disposal of municipal waste. The population is charged on a per capita basis. The rate is 700 manats per month and covers the collection, transport and disposal of waste. The charge collection rate is quite high, at around 95%.

International agreements and activities

Azerbaijan takes an active part in the activities under the 1989 Basel Convention on the Control of Transboundary Movements of Hazardous Wastes and their Disposal, to which Azerbaijan acceded in June 2001. The Hazardous Waste Management Agency is involved in the implementation of the Convention. At present a report on the activities concerning hazardous waste management and its movement is being prepared. The report will be submitted to the secretariat of the Convention and presented to its Parties (see also chapter 4 on international cooperation).

6.8   Conclusions and Recommendations

Azerbaijan recognizes the importance of waste management but has not yet been in a position to allocate sufficient resources to it. Although, at present, many industries do not work at all or function at only 10-30% of their capacity, the problem of accumulated industrial and hazardous waste remains critical. Similarly, the situation with disposal of municipal waste needs to be addressed. In both cases, regular and reliable data need to be collected.

Oil and gas production is the main contributor to GDP in Azerbaijan, but it has also led to serious contamination of land, especially in the Absheron Peninsula, and of the Caspian Sea. A number of positive changes are however taking place to improve the situation. A Hazardous Waste Management Agency was established, with important new responsibilities regarding the generation and disposal of hazardous waste, including mercury sludge. The rehabilitation of oil-contaminated sites has started and will continue. Nevertheless, more still needs to be done to solve all the waste management problems.

A draft national hazardous waste management strategy has recently been developed. It is now in the process of review and adoption by the Government. The strategy is a comprehensive and important document, which could be used to improve the waste management system. Among other issues, the draft strategy would provide for an overhaul of the legislative framework for waste management and for the mechanisms needed for implementation.

The work on the environmentally sound management of hazardous waste has started, but the overall system of hazardous waste management still requires further development and improvement.

The management of household waste also continues to pose a number of serious challenges. There are no landfills that meet sanitary standards and norms. Among other problems, landfills are not guarded, waste is burned and there is no protective lining to prevent leaching of substances into ground water.

Recommendation 6.1:
(a) The Ministry of Ecology and Natural Resources, in cooperation with industry, should further develop the hazardous waste management system, which is currently in its initial stages. It should also include development and improvement of its infrastructure (testing procedures, laboratory practices, standard analytical methods for defining waste composition as well as technical guidelines on waste handling);
(b) The Government should adopt the draft national hazardous waste management strategy as soon as possible. The Ministry of Ecology and Natural Resources should facilitate this process, as appropriate;
(c) The Ministry of Ecology and Natural Resources, in cooperation with other relevant ministries, should develop and implement a new comprehensive law on waste management with relevant regulations and norms.

Information is essential for good policy. At the moment, statistics for industrial waste do not include all waste and do not reflect the real situation. An environmental audit of functioning enterprises as well as abandoned industrial sites is needed in order to identify all sources of industrial (non-hazardous) and hazardous waste. Similarly, an inventory of land contaminated by hazardous waste is essential before a systematic programme of clean up and reclamation is undertaken. These should be high priorities for the Government.

The Ministry of Ecology and Natural Resources provides information about mercury sludge and obsolete pesticides, and this can be found in the statistical reports. More efforts could be made to make this information accessible to the public, through, for example, issuing bulletins or information sheets, particularly as the issue relates to human health and the environment.

Recommendation 6.2:
The Ministry of Ecology and Natural Resources, in cooperation with the Ministry of Economic Development and industrial enterprises, should:
* *Conduct environmental audits of functioning industrial enterprises;*

* *Draw up an inventory of abandoned industrial sites and create a database of all industrial waste;*
* *Prepare and implement an action plan for the rehabilitation of oil-contaminated sites by mechanical and/or biological method, including a mechanism for financing;*
* *Make information concerning the threats to health posed by hazardous waste disposal sites readily available to the public.*
* *Ensure that the sites are fully contained and inaccessible to the public.*

A TACIS grant has recently provided significant funding to upgrade the IZOTOP centre, which has served as the only controlled landfill for radioactive waste in Azerbaijan since 1963. It is important that the facility should fully follow international good practices for storing radioactive waste, and that it should be continuously maintained.

Recommendation 6.3:
The executive power of Baku, in cooperation with other institutions involved in radioactive waste management, should:
* *Draw up an inventory of all radioactive sources;*
* *Rehabilitate the IZOTOP centre facility to meet international norms and standards for the environmentally sound disposal of radioactive waste. The long-term sustainability of the IZOTOP centre and maintenance of the plant should be ensured through fees to be charged to private enterprises that use this service. Public entities are expected to receive the service free of charge.*

Obsolete pesticides are stored at an uncontrolled site that is in a very poor condition. Some concrete containers are unsealed, which results in the discharge of volatile components of pesticides into the atmosphere, as well as in mixing of rainwater with pesticides and its subsequent leaching into groundwater. These pesticides contain a number of chemicals, including DDT. An environmental impact assessment of the site has never been conducted, and there is no monitoring system. Pesticides are no longer produced in Azerbaijan.

Recommendation 6.4:
The Ministry of Ecology and Natural Resources, in cooperation with the Ministry of Agriculture, should conduct an environmental impact assessment of the pesticides storage facility and begin its rehabilitation, ensuring that the storage

facility is contained, that there is no leaching into the soil and groundwater and that it can withstand all weather conditions;

The system for municipal waste collection and disposal works well in Baku city, but there is no separation of municipal waste, except for glass bottles and bread. In cities and towns outside Baku, municipal waste is collected and transported to landfills, without any separation or treatment. Existing landfills do not meet sanitary requirements according to European standards and norms. As a result, there are several potential adverse effects, including penetration of groundwater with heavy metals, toxic organic chemicals and hazardous chemicals as well as contamination of the air in the vicinity of the landfills.

Recommendation 6.5
Municipalities, in cooperation with the Ministry of Ecology and Natural Resources, should:
- *Organise awareness-raising campaigns among the population to encourage them to separate recyclable waste;*
- *Construct facilities for the collection and reprocessing of this waste;*
- *On a step-by-step basis, construct new sanitary landfills for disposal of municipal waste on the basis of environmental impact assessments;*
- *In the long-term, construct incineration facilities for municipal waste in Baku.*

Chapter 7

WATER MANAGEMENT

7.1 Water Resources

Availability

Azerbaijan is poor in terms of available water resources. A large part of the country, notably the Kura-Araz lowlands and the Absheron peninsula, has a significant water deficit due to low precipitation and high evaporation. This part, which houses more than 70% of the population, is completely dependent on irrigation for its agricultural production.

The Kura is the main river in the region. Its source is in Turkey, and the river and its tributaries flow through Georgia, Armenia, the Islamic Republic of Iran and Azerbaijan before discharging into the Caspian Sea. The Kura river basin covers 80% of the territory of Azerbaijan. The general absence of waste-water treatment in the river basin with a population of 11 million and economic activities results in very low water quality in the region and especially in Azerbaijan. This is a major problem, with Azerbaijan being dependant on the Kura river for more than 70% of its drinking water supply.

The coastal areas to the north and south of the Kura basin are drained by smaller rivers to the Caspian Sea. One of these is the Samur river, on the border with Dagestan (Russian Federation), which plays an important role for water supply to Baku and for irrigation of Absheron.

Groundwater provides only about 5% of the total water abstraction. It plays an important role, however, for irrigation and water supply to provincial towns and rural areas, especially in the foothills of the mountain ranges.

Climate and water balance

The climatic zones range from subtropical to temperate and alpine climates. Dry subtropical climate is typical for the Kura-Araz lowlands and

Absheron peninsula. Temperate climate dominates the mountain slopes of the Greater and Lesser Caucasus Mountains covered by forests. The high areas of these mountains are alpine environments.

Rainfall varies from the south coast of Absheron, where it is less than 200 mm per year, over less than 400 mm in most of the Kura river basin, to 600-900 mm in the foothills and mountainous zones, 1,000-1,300 mm on the southern slope of the Greater Caucasus, and 1,200-1,400 mm/year in the southern Lenkeran lowlands. Evaporation in the Azerbaijani part of the Kura river basin is as high as 93% of annual precipitation (compared to 61% in Armenia and 50% in Georgia). This is an indication of the serious water balance situation in Azerbaijan compared to its neighbours. The total water balance in the Kura basin is shown in table 7.1 below.

Quantity and quality of groundwater

Groundwater resources are mainly restricted to the foothills and intermountain plains of the Greater Caucasus, Lesser Caucasus, Nakhchivan and Talysh (see map). The total groundwater resource is estimated at 24 million m^3 per day (8.8 km^3 per year). At present 5 million m^3/day or 20% is exploited. Groundwater is mainly used for irrigation (78%), 3% is used by industry and 19% for water supply to provincial towns and rural areas. Groundwater data are given in table 7.2 below.

Table 7.1: Indicative water balance in the Kura basin

			(km³)
	Azerbaijan	**Armenia**	**Georgia**
Precipitation	31	18	26
Effective evaporation	-29	-11	-13
River inflow	15	1	1
River outflow	-18	-8	-12
Underground inflow	3	1	1
Underground outflow	-2	-1	-3

Source: Tacis. Joint River Management Programme. January 2003.

Table 7.2: Groundwater

Reservoir	Estimated capacity, m³/day x 1000	Typical borehole depth, m	Typical borehole yield, l/s	Domestic usage m³/day x 1000	Irrigation and industrial usage, m³/day x 1000	Total usage, m³/day x 1000
Total	**23,764**			**969.3**	**4,094.3**	**5,063.6**
Mountain zone of Greater Caucasus	1,009	100-150	1-5	12.3	7.3	19.6
Mountain zone of Lesser Caucasus	989	100-150	1-5	8.0	8.8	16.8
Absheron peninsula	242	20-70	0,5-2	19.3	33.6	52.9
Samur-Gusar pre-mountain plains	3,471	150-200	15-30	391.2	41.3	432.5
Sheky-Zakatalar pre-mountain plains	3,822	100-150	20-40	72.4	191.7	264.1
Ganja-Gasakh pre-mountain plains	4,219	120-200	15-25	211.8	1,121.0	1,332.8
Shirvan pre-mountain plains	518	80-120	2-8	28.9	15.3	44.2
Karabakh-Mil pre-mountain plains	7,910	120-150	15-20	110.7	2,392.3	2,503.0
Mugan plains	130	50-120	1-5	3.6	40.9	44.5
Jebrail pre-mountain plains	235	150-200	10-15	1.1	6.7	7.8
Lenkeran plains	209	20-120	2-10	34.8	97.7	132.5
Nakhchivan pre-mountain plains	989	120-150	3-10	75.2	137.7	212.9

Source: Ministry of Ecology and Natural Resources. January 2003.

In some areas groundwater is shallow and poorly protected, e.g. Absheron peninsula. Groundwater in the lowlands often has a high salt content (nitrites up to 1,2 mg/l and nitrates up to 75 mg/l) due primarily to poor drainage and, to some extent, to the use of fertilizers. On the Absheron peninsula the mineral oil content is high in some areas (up to 50 mg/l). Industrial pollution is found in Sumgayit (heavy metals) and Ganja (aluminium up to 3,5 mg/l and iron up to 50 mg/l). Bacteriological pollution of the upper aquifers is observed in the irrigated areas, towns and on cattle farms.

Quality and quantity of surface waters

The Kura is the major river system accounting for approximately 90% of surface water resources in Azerbaijan. The Kura river rises in Turkey and passes through Georgia before entering Azerbaijan on its 1,500 km journey to the Caspian Sea (900 km in Azerbaijan). The Araz river, a major tributary to the Kura, also rises in Turkey. It flows along the border between Armenia and Iran before joining the Kura. The Kura basin occupies 68,900 km² of Azerbaijan or 80% of its territory.

The remaining 10% of the surface water resources is made up of a number of rivers originating from the Greater Caucasus, the Lesser Caucasus and the Talysh, and flowing directly towards the Caspian Sea. Many of these rivers

are however not perennial or disappear underground before reaching the sea. The main rivers are listed in table 7.3 below.

In Azerbaijan there are 23 main reservoirs, of which only 3 have a volume over 1 km³. The Mingechevir reservoir on the Kura river is the biggest, with a capacity of 15.7 km³. The water is used for power generation and for irrigation. The Jerianbatan reservoir (0.2 km³), north of Baku, is fed by the Samur river on the border with the Russian Federation through a 180-km-long channel. The reservoir is important for the water supply to Baku and for irrigation.

Table 7.3: Main rivers in Azerbaijan

River name	Annual runoff (million m³/year)	% of total runoff
Total	**11,962**	**100.0**
Kura and tributaries	10,911	91.1
Lenkeran	319	2.7
Kudialchay	184	1.5
Velvelichay	126	1.0
Kursarchay	119	1.0
Velyashchay	117	1.0
Tangeru	109	0.9
Karachay	41	0.5
Charakjukchay	25	0.2
Istisuchay	11	0.1

Source: Ministry of Ecology and Natural Resources. January 2003.

Reservoir and dam construction also serves flood regulation, and the Mingechevir has improved the situation in this respect in the Kura lowlands. Downstream of the confluence of the Araz river, however, floods frequently occur due to a combination of increased water level in the Caspian Sea and sedimentation in the river bed. Emergency work on the Kura dykes in May and June 2003 mitigated the impact of flooding in the Salyan and Nefchala areas. Deforestation in the upper part of the river catchment areas has led to poor soil protection with damaging mud slides as a result. Flash floods are frequent.

The Kura river system is organically and bacteriologically polluted by the discharge of poorly treated or untreated waste water from the 11 million people living in the catchment area. Due to the collapse of many industries in the early 1990s, pollution has decreased considerably. A number of polluting activities, however, still exist, notably mining, metallurgical and chemical industries. The major pollutants are heavy metals (Cu, Zn, Cd) from mining and the leather industry, and ammonia and nitrates from the fertilizer industry. Concentrations exceed norms up to nine times. Phenols exceed the norms six times and mineral oil, two to three times. Pesticides, especially organo-chlorine pesticides such as DDT, also constitute a problem. DDT is no longer used and the source of this pollution has not been fully investigated; it may be old stock or illegally produced or imported products. (See chapter 10, on Land use, agriculture and desertification.)

Deforestation and overgrazing have led to erosion causing high turbidity of river water. The Araz river is claimed to be one of the most turbid in the world. High turbidity increases the cost of drinking water

production. This is important since the main part of the drinking water for Baku and the rest of the population on the Absheron peninsula is abstracted from the Kura river after the confluence with the Araz. The quality problems are especially serious for Azerbaijan since it is the end recipient in the Kura system.

7.2 Water Uses and Pressures on Water Resources

Water use

The State Committee of Amelioration and Water Management receives annual water use reports from all water users (1,860 users in 2002). Total water use in 2002 was 10 billion m^3, with 6% used for domestic water supply, 20% for industrial water supply and 42% for irrigation. Thirty-two per cent of the water abstracted was lost in the systems. Compared to nine years earlier (1993), total water abstraction had decreased, by 38%. Irrigation had gone down by 49% and industrial uses by 43%. Domestic uses, however, had increased from 2.5 to 5%, and water losses from 26% in 1993 to 33% in 2002. Surface water constitutes 95% of the water resources used. Per capita abstraction was 2,149 m^3 in 1993. In 2002 this had decreased to 1,256 m^3 per capita, or a reduction of 42%. Water use data are given in table 7.4.

Household use of water

Water-supply coverage in Azerbaijan is estimated at 50%, which is relatively high compared with other countries with similar income levels. The rural coverage is probably higher than indicated in table 7.5 below, as water provided by rural communities relying on wells or irrigation canals is not included in the official statistics.

Table 7.4: Water abstraction and use

(million m^3/year)

	1993	1994	1995	1996	1997	1998	2002
Total abstraction	**16,344**	**14,631**	**13,970**	**13,462**	**12,512**	**10,235**	**10,075**
- surface water	15,156	13,118	12,820	12,475	11,414	9,554	9,530
- groundwater	1,188	1,513	1,150	987	1,098	681	545
Per capita abstraction, m3	2,149	1,923	1,837	1,753	1,613	1,307	1,256
Domestic use	390	368	327	277	222	264	503
Industrial use	3,459	2,323	2,173	2,225	2,132	2,293	1,977
Irrigation	8,222	7,996	7,668	7,047	6,397	4,482	4,169
Cattle and other use	78	99	55	383	284	254	105
Water losses	4,195	3,855	3,747	3,530	3,477	2,941	3,321

Source: The Committee of Amelioration and Water Management. January 2003.

Table 7.5: Water-supply coverage

	Connected to water supply, %
Baku	95
Sumgayit, Ganja	95
Secondary cities	83
Rural areas	11

Source: World Bank. Azerbaijan Water Supply and Sanitation Sector Review and Strategy. 2000.

The reliability and safety of the services are a major problem due to the poor state of repair, lack of maintenance and insufficient resources available for operations. On average water was available to individual households in the Greater Baku area about 22 days per month, four hours per day (household surveys in 1995 and 1998). A US$95 million rehabilitation project recently completed on the Greater Baku water-supply system has improved the headworks (reservoirs and treatment plants), but much higher investments in pumping stations and the pipe network are still needed (estimated at US$1 billion). To cope with this situation, 53% of households in the Baku area have invested in their own overhead storage tanks and 12% in water pumps.

Per capita water use is very high in Azerbaijan. Individual water meters are installed only in very few cases, and it is therefore impossible to determine the actual per capita consumption. Estimates made as part of the Master Plan for Greater Baku show an average consumption in Baku of 580 litres per capita per day. In Baku 5,000 meters were installed as part of the rehabilitation project, and water use in metered households is claimed to have dropped from 600 to 400 litres per capita. The same trend is seen in Imishly, a secondary provincial town, where water meters have been installed as part of a privatization project. For a well-run water utility with household meters, per capita consumption of 150 to 200 litres per day would be expected. The high water consumption figure is a result of several factors: the poor state of the transmission and distribution pipe network, the poor state of installations in the home and the absence of metering.

Water for Baku, Sumgayit and the Absheron peninsula is provided by the Absheron Regional Water Company, a public-owned joint-stock company. Three water sources are used: the Shollar and Hachmas springs, Jerianbatan reservoir fed by the Samur river, and the Kura river. Water quality

of the spring sources is high (only chlorination needed). The Samur river water is of reasonable quality after natural cleaning in the Jerianbatan reservoir. Full treatment is still required, however. This is also the case for the heavily polluted water from the Kura river. For this reason the intake on the Kura is used to fill the gap after taking the maximum quantity from the springs and the Jerianbatan reservoir.

Sixty cities and towns with populations ranging from 52,200 (Samukh) to 301,400 (Ganja) have water-supply systems. Only 14 of these towns rely fully or partly on surface water. Groundwater is assumed to provide 90% of their total water supply. The capacity of these systems is heavily used, resulting in per capita uses as low as 30 to 100 litres per day. In rural areas groundwater wells are common but irrigation canals are also frequently used as a source of water for domestic use.

The excessive water losses in the Greater Baku area indicate that the existing water sources would not need to be expanded to serve the region adequately. However, the Greater Baku area relies on water from the Samur river that may be subject to changes to the international allocation agreement with the Russian Federation, and from sources that are also used for irrigation which in the future may be a cause of conflict among users, particularly in the dry season.

The quality of drinking water in the networks does not meet international standards. The limited data available indicate that bacteriological requirements are exceeded and minimum residual chlorine content is not met. There are several reasons for this. First, surface water is highly polluted, making treatment complicated and costly; second, water treatment is insufficient (outdated facilities and lack of chemicals, e.g. chlorine); and third, the water distribution network is leaking and operated intermittently, which causes contaminated groundwater to enter the water network during periods of low pressure.

Waste water

The waste-water network in Baku serves about 72% of the city but only about 50% of the waste water is treated; 90% biologically and 10% mechanically. In other urban areas in the country, the coverage drops to 32%. There are waste-water treatment plants in 16 cities and towns; most are partly or completely out of operation. In rural areas, on-site sanitation is used, primarily latrines.

Table 7.6: Total loads of pollutants discharged to the Caspian Sea from Azerbaijan in 1998

Source	Flow 10^6 m^3/y	%of total flow	BOD t/y	%of total load	Inorg. N t/y	%of total load	Phosphorus t/y	%of total load	Hydro-carbons t/year	%of total load
Total	12,619	100	61,634	100	57,764	100	3,622	100	15,764	100
Rivers	11,960	95	17,242	28	44,673	77	153	4	500	3
Municipal	485	4	38,500	62	12,500	22	3,250	90	9,408	60
Industrial	174	1	5,892	10	591	1	219	6	5,856	37

Source: Tacis. Assessment of Pollution Control Measures. 2000.

The condition of waste-water facilities in Azerbaijan is generally very poor. Lack of maintenance for more than a decade, the excessive flows due to leakage and infiltration, and the low standard of construction and materials are the main reasons for this. Discharges of insufficiently pretreated harmful industrial waste water into municipal sewer systems impair the efficiency of the waste-water treatment plants not designed to deal with these loads.

The total load of pollutants from Azerbaijan to the Caspian Sea is summarized in table 7.6. The table shows that 72% of biological oxygen demand (BOD), 96% of phosphorus and 97% of hydrocarbons come from domestic and industrial waste water. Nitrogen stems from irrigated areas and is primarily (77%) discharged through rivers.

Table 7.7 shows that Baku accounts for approximately 75% of the pollution load from domestic waste water on the Caspian Sea.

The waste-water system of Baku is operated by the executive power of Baku, organized in a separate department called Bakkanalizatsiya. The system, which consists of a collection network and five treatment plants, is in dire need of rehabilitation. A feasibility study conducted in 2000 estimates the investment needed at US$1.3 billion.

Agriculture

Azerbaijan is dependent on irrigation for most of its agricultural production. In 2002, 4.2 billion m^3 were used for irrigation. This constitutes 41% of the total water abstraction, making irrigation the largest water user. The total area with installed irrigation is 1.45 million hectares (nearly 85% of the cultivated area). Water use for irrigation has dropped significantly since Azerbaijan's independence. In 1993 irrigation used 8.2 million m^3, 95% more than in 2002.

The irrigation infrastructure suffers from a number of problems, including:

- Deterioration of infrastructure and pumping equipment due to insufficient maintenance;
- High reliance on pumped irrigation (over 500,000 ha), which in many instances would make agriculture uneconomic if the energy were valued at its real cost;
- Negligible contribution from users to operation and maintenance expenses;
- Inefficient water distribution and application.

Recent efforts to improve the situation have led to the establishment of institutional mechanisms for the collection and use of water charges and the transfer of responsibility to water users.

Table 7.7: Loads of domestic waste water, treated or untreated, discharged directly to the Caspian Sea from Azerbaijan in 1998

Source	Flow 10^6 m^3/year	%	BOD Tons/year	%	Total-N Tons/year	%	Total-P Tons/year	%	E. coli 10^{15} c./year	%
Total Azerbaijan	485	100	37,500	100	12,450	100	3,250	100	2,740	100
Baku	378	78	24,500	65	9,250	74	2,400	74	1,700	62
Sumgayit	60	12	7,500	20	1,800	14	480	15	600	22
Other	47	10	5,500	15	1,400	12	370	11	440	16

Source: Tacis. Assessment of Pollution Control Measures. 2000.

It is estimated that 40-45% of the irrigation infrastructure is in need of rehabilitation. The inefficient use of water and the high water losses in irrigation are major problems in relation to water resources and soil.

Industry

Nearly 70% of Azerbaijan's industrial complex is concentrated on the Absheron peninsula with the two largest industrial centres located in Baku and Sumgayit. After the collapse of the Soviet Union the industries in Azerbaijan lost the bulk of their market. This has led to low production or the outright closure of many factories, resulting in a lower discharge of industrial waste water. Industrial water use has gone down from 3,418 million m^3 in 1990 to 1,977 million m^3 in 2002, a 42% reduction.

Water for industrial use is often supplied by the public system. However, some industries have their own system based on surface water or groundwater.

A study prepared by TACIS in 2000 concluded that, out of 30 operational industries in Baku, 1 discharged waste water to the municipal system and 29 had their own systems discharging to the Caspian Sea. Three industries had biological treatment plants, 20 had mechanical treatment and 7 had no treatment. Three industrial enterprises have significant waste-water production, i.e. two oil refineries and one power plant. They all have treatment plants and discharge directly. Despite this one of the municipal waste-water treatment plants (Zikh) located near a refinery receives sewage with significant oil content. In Sumgayit only 10 of 19 industrial enterprises were operating; 4 had biological treatment, 2 mechanical and 4 had no treatment. Six industries had significant waste-water production: two power plants and four chemical industries. There are no recent data on industrial water pollution. It is known, however, that significant quantities of heavy metals and organic pollutants are discharged by industry.

Outside Baku and Sumgayit, as few as 33 operating industries are registered, the majority connected to agricultural production. Overall, their waste-water discharge is insignificant, although local problems may exist where the capacity of the recipient is limited, e.g. a small river.

7.3   Policy Objectives and Management

Policy framework

The Water Code sets the basis for water management in Azerbaijan and gives the following main principles for use and protection:

- Economic development and environmental protection;
- Provision of the population with quality water;
- Water management should be based on river basins; and
- Water protection functions should be separate from water use and water industry functions.

Meeting these objectives poses a big challenge. The present coverage with quality drinking water is 50%. Water management is based on administrative units rather than river basins and there is very little coordination among the countries in the main river basin, the Kura. Although the creation of the Ministry of Ecology and Natural Resources has provided a clearer management structure in the water sector, overlapping functions and responsibilities remain.

A workshop in February 2003 with participants from various agencies and other entities addressing water resources in Azerbaijan concluded that the preparation of a national integrated water use and water protection strategy should be given top priority. The strategy should be based on an integrated river basin management approach rather than administrative territorial water management.

Legal framework

The legal framework for the water sector consists of the following laws:

- The Water Code (1997);
- The Law on Water Supply and Waste Water (1999);
- The Law on Amelioration and Irrigation (1996); and
- The Law on Environmental Protection (1999).

The Water Code sets the basis for water management (see section on the policy framework). The Law on Water Supply and Waste Water sets the legal framework for this sector. Its important features are:

- Responsibility for providing water and sewerage services is given to enterprises;

- The management and operation of water-supply enterprises are to be regulated by a contract with the municipalities they serve;

- Enterprises now have the right to cut services to consumers in case of non-payment or illegal connections; and

- Metering of water supply is recognized as the main method for charging for water services.

The Law on Amelioration and Irrigation regulates the planning, design, construction and operation of amelioration and irrigation systems. It states that design and construction activities require special permits (licences). Systems have to be certified with technical passports.

The Law on Environmental Protection identifies the legal, economic and social bases of environmental protection. It governs the use of natural resources (e.g. water) and protection against pollution (e.g. domestic and industrial). The Law also sets the basis for economic mechanisms, e.g. payment for the use of natural resources, payment for the disposal of domestic and industrial waste, and economic incentives for environmental protection. The environmental requirements and approval procedures in connection with the construction or reconstruction of municipal and industrial facilities are defined in the Law. It includes a very detailed description of the ecological expertise to be conducted.

These laws are supplemented by a large number of decrees issued by the President and decisions issued by the Cabinet of Ministers.

Regulatory instruments, standards and norms

The State Committee of Amelioration and Water Management is responsible for issuing water abstraction permits for surface water. It is also responsible for imposing payments for water use. Since 1997 water that is used in agricultural purposes is chargeable. Rates for fees for water use were changed in June 2003. The fee is charged for technical-operational costs and not for use of water as a natural resource

The National Geological Exploration Service, a department of the Ministry of Ecology and Natural Resources, is responsible for regulating and controlling abstraction of groundwater.

The Ministry of Ecology and Natural Resources issues waste-water discharge permits, which are valid for 3 to 5 years. Its regional offices control and enforce discharge permit conditions.

The Ministry of Health, through its Centre for Epidemiology and Hygiene, is responsible for monitoring drinking-water quality.

Norms and standards

The design of water-supply and waste-water infrastructure is based on the building codes developed during Soviet times known as SNIP standards. These standards set high consumption rates, usually 400 litres per capita per day (lcd). Furthermore, they require a high level of supply safety (e.g. duplication of main pipelines and high storage capacity). The result is oversized and therefore costly systems. The standards lead not only to over production but also to wasteful consumption practices.

The quality of drinking water must comply with World Health Organization (WHO) standards. Most water-supply systems have difficulties in meeting these standards. While they are certainly a commendable goal, it might be better to achieve a minimum acceptable level of safe drinking water. Additionally, the large number of parameters and the high sampling frequency would require very sophisticated equipment and many staff. The result is little follow-up on water quality and compliance.

Azerbaijan's effluent standards for waste-water treatment plants are among the most demanding in the world. For example, the standard for BOD is set at 3 mg/l for fishing water and 6 mg/l for other waters. Standards for other pollutants are similarly strict. In comparison, the EU standard for BOD is 25 mg/l. In Azerbaijan waste water is chlorinated. This is unnecessary and actually gives rise to harmful organochlorine compounds in the effluent. The strict requirements for waste-water treatment naturally increase the construction and operation cost of waste-water systems.

Economic instruments

Charges are levied on surface and groundwater abstraction as well as on use. The charges were introduced in 1992 to stimulate the rational and integrated use of water and to raise funds for water protection. However, the rates have not been adjusted to take account of the high inflation in the 1990s.

Charges on waste-water discharge were also introduced in 1992. The rates are again very low and so is the collection rate, which weakens the effectiveness of the charge system. Furthermore, the near collapse of the charge system has eliminated the primary source of financing for monitoring and enforcement.

Consumers are charged for water-supply and waste-water services. The tariff systems are based on heavy cross-subsidies from industry to domestic users. The rates for domestic users are very low. The Absheron Regional Water Company charges 185 manats/m^3 or US$ 0.04 /m^3). The rates for budget organizations and industries are 800 manats/m^3 and 2,200 manats/m^3 respectively. At the same time the Company's collection rate is low: 80%. Although the rates are based on consumption, there are few water meters (only 1.1% of the Company's domestic customers have water meters), so in fact a flat rate of 12 m^3 per person per month is used. The rate of metering is higher for industries; 52% of the Company's industrial customers have meters. The revenue collected by the Company does not cover its operating cost and as a result the Company is itself in arrears, especially with its energy bills.

In the provincial town of Imishly, where a German company has signed a 10-year management contract (2000 to 2010) for the water-supply system, the rates per m^3 have been set at 1,000 manats for domestic use, 3,000 manats for budget organizations and 5,000 manats for commercial use. Water use is metered, and the revenue covers the investment and operating costs. The project is meant as a pilot for provincial towns and the company is considering similar projects in two other towns.

Waste-water services are charged to the users. Bakkanalizatsiya, the Baku municipal department responsible for waste-water services, has set the following rates: 40 manats/m^3 for households, 444 manats/m^3 for industries, 354 manats/m^3 for budget organizations and 2,360 manats/m^3 for commercial use. Waste-water quantities are based on water use (either metered or 12 m^3 per person per month). The revenue collected covers only 40-50% of the operating costs and Bakkanalizatsiya, too, is in arrears with its payments to other public entities.

Institutional framework

The following ministries and institutions are involved in water management:
- The Ministry of Ecology and Natural Resources;
- The Ministry of Health;
- The State Committee of Amelioration and Water Management;
- The State Committee of Architecture and Construction;
- The Absheron Regional Water Company; and
- The executive power of Baku.

The Ministry of Ecology and Natural Resources has overall responsibility for the conservation of water resources and the prevention of pollution. In the process of establishing the Ministry, a number of State committees and other organizations were transferred and became departments in the new Ministry, and several of these are involved in the water sector: the National Geological Exploration Service is responsible for the regulation and control of groundwater abstraction, the National Hydrometeorological Service is responsible for surface water monitoring, operating of 99 hydromet stations and publishing the results in an annual report. Water quality monitoring in surface water and groundwater is the responsibility of the National Monitoring Service.

The Ministry of Health, through its Centre for Epidemiology and Hygiene, is responsible for setting drinking-water standards and monitoring its quality.

The State Committee of Amelioration and Water Management is responsible for monitoring water use and issues annual reports based on information from water users (1860 users in 2002). The Committee's activities focus on irrigation, for which it sets water-use norms and handles user relations, for example. It is also responsible for establishing the charges for water use. In addition, it is in charge of land improvement on irrigated land, and the operation and maintenance of the irrigation infrastructure.

The State Committee of Architecture and Construction is responsible, among other things, for water-supply and waste-water services outside the supply area of the Absheron Regional Water

Company. In urban areas the Committee works through the municipal water and waste-water enterprises (*vodokanals*). The Committee is meant to have an advisory and monitoring role; however, due to the relative weakness of the municipalities, the Committee has taken on a managing role too. Rural water supply also falls under its responsibility.

The Absheron Regional Water Company is a joint-stock company responsible for the treatment and distribution of water to Greater Baku, including bulk water supply to the *vodokanals* of Absheron and Sumgayit. The Company's share capital is fully owned by the State.

The waste-water collection and treatment services in Baku are the responsibility of Bakkanalizasiya, a department under the executive power of Baku. The plan is to merge the Absheron Regional Water Company and Bakkanalizasiya into a joint-stock company and privatize it.

Role of intersectoral and international cooperation in water management

The Organization for Security and Co-operation in Europe (OSCE) facilitates regular donor meetings on water sector issues. In May 2003 OSCE and the United States Agency for International Development (USAID) organized a regional workshop in Georgia on the priorities in the water sector. Prior to this, national workshops had been conducted in Armenia, Azerbaijan and Georgia. Although only bilateral cooperation is possible owing to the present political situation, fruitful discussions were held at the administrative level.

The TACIS-financed Joint River Management Programme includes the Kura river. Armenia, Azerbaijan and Georgia participate. The Programme, which started in 2002 and will run for two years, will help to prevent, control and reduce transboundary pollution caused by the water quality of the Kura river. The project focuses on issues such as water quality and quantity monitoring, transboundary pollution reduction, pollution warning systems, legal and regulatory improvements, raising public awareness, and transboundary agreements.

The World Bank, in cooperation with Azerbaijan's Committee for Housing and Communal Property and the Absheron Regional Water Company, prepared the National Water Supply and Sanitation

Sector Review and Strategy in 2000. Its proposed strategy encompasses four key reforms: (i) institutional and governance; (ii) financial; (iii) technical; and (iv) service. The institutional strategy proposes that the water-supply and waste-water systems of Baku, Sumgayit and Absheron be merged into one enterprise operated by an international utility operator. For medium and small urban areas (population >5,000) the proposal is to restructure water-supply and waste-water utilities and decentralize responsibility supported by regional service units. In rural areas (communities <5,000) with relatively simple water-supply systems, the vision is that communities own, manage and help finance their facilities. The financial strategy encompasses rigorous bill collection, tariff levels for full cost recovery, revision of cross-subsidies and improved financial management. The technical reform includes installation of water meters, reduced network leakage, increased energy efficiency, pretreatment of industrial waste water, and appropriate and low-cost solutions for rural areas. The service reforms include the following elements: improved service to the poor, appropriate and affordable standards and level of service, and development of human resources.

As part of the implementation of the Water Supply and Sanitation Sector Strategy a number of World Bank financed projects are underway. One project will propose a regulatory framework for the utility sectors, based on either a common framework or individual regulations for water, gas and electricity. Another project will develop a framework and policy for the water sector, and a third project will develop a privatization strategy for the combined water-supply and waste-water company of Greater Baku.

The World Bank and the European Bank for Reconstruction and Development (EBRD) are financing the Greater Baku Water Supply Rehabilitation Project. The Project, which started in 1996, made emergency short-term improvements in the water-supply system in order to restore the provision of water supply to Baku, i.e. rehabilitation of the water-treatment plants on the Kura river and the Jerianbatan reservoir, and of intake and distribution systems. In November 2000 a strong earthquake hit Azerbaijan and badly damaged the water-treatment plants and the distribution network in Baku. A supplemental credit has been given to repair the damages.

Kreditanstalt für Wiederaufbau (KfW), Germany's development bank, is financing a project in Imishly, a provincial town with approximately 40,000 inhabitants. The project started in 2001 with the rehabilitation of the water-supply system and the installation of water meters. A ten-year management contract for the Imishly water utility has been signed with a German utility company.

USAID finances the South Caucasus Water Management Project covering the Kura basin. The Project's goals are to improve water quality and quantity monitoring, to improve capacity to analyse and to implement watershed management pilot projects.

The Asian Development Bank finances the Secondary Towns Water Project. The Project, which started in 2002, includes the rehabilitation of existing water-supply systems. So far three towns have been selected: Goychay, Aghdash and Nakhchivan.

Protection of transboundary rivers

Azerbaijan is Party to three agreements with its neighbours on transboundary rivers: with the Islamic Republic of Iran on the Araz river, with Georgia on Gandar Lake and with the Russian Federation on the Samur river. The authorities in Dagestan have requested a renegotiation of the Samur agreement in order to get a larger share of the water, reflecting the division of catchment area between Dagestan and Azerbaijan.

No agreement exists on the Kura river, the most important transboundary river in the region. The Kura and Araz rivers are by far the most important water resources for Azerbaijan. The poor water quality of the river system is a consequence of pollution from cities, agriculture and mining in Armenia, Georgia and Azerbaijan. The solution to these problems requires coordinated national and international efforts. The Republic of Azerbiajan is Party to the Convention on the Protection and Use of Transboundary Waters and International Lakes, but not all neighbouring countries are Parties, making solution of this problem more complex.

Several of the projects mentioned above are transboundary with the aim of improving regional cooperation. Important elements of any future agreement, such as improvements in

water quality and quantity monitoring and data handling, are included in the projects.

The National Programme on Environmentally Sustainable Socio-economic Development, launched by the President in February 2003, stresses the need for increased regional and international cooperation to protect transboundary rivers from pollution and ensuring the effective use of water by riparian countries.

7.4   Conclusions and recommendations

The water sector of Azerbaijan faces enormous problems. Adverse climatic conditions with low precipitation and high evaporation cause widespread water shortages. Poor-quality water-supply and irrigation networks cause very high losses. Payment systems are not based on actual water use and therefore give no incentive to save water. Water resources are polluted owing to the lack of waste-water treatment plants in Azerbaijan and neighbouring countries. The quality of drinking water does not meet the required standards. Owing to inflation, economic instruments such as abstraction charges and user fees have become meaningless.

The Government of Azerbaijan has taken a number of steps to reverse this negative situation. The most recent of these is the National Programme on Environmentally Sustainable Socio-economic Development launched in February 2003. It includes a number of specific actions aimed at improving the situation before 2010. The following recommendations in most cases coincide with the Government's plans and should therefore be considered as support to its efforts.

Many of the problems mentioned above are related to the lack of efficient cooperation among the stakeholders in the water sector. The creation of the Ministry of Ecology and Natural Resources was a clear improvement in this respect. The State Committee of Amelioration and Water Management focuses on water regulation and irrigation. The water-supply interests are defended by the Absheron Regional Water Company and the State Committee of Architecture and Construction. Waste-water management involves a number of entities: Baku and Sumgayit executive powers, the State Committee of Architecture and Construction and industries. Others with an interest in water include:

hydropower-generation plants, farmers' associations and domestic water users. All these stakeholders should be involved in establishing a common vision for the water sector. The basis should be a river basin approach rather than an administrative, territorial approach.

The deteriorating water quality of the Kura river is a major problem for Azerbaijan. It cannot be solved without involving Armenia and Georgia. Although multilateral negotiations cannot take place at the political level at present, initiatives by international organizations have made technical cooperation possible. This will be very important for the preparation of political discussions once this will again be possible.

Recommendation 7.1:
The Ministry of Ecology and Natural Resources and the State Committee of Amelioration and Water Management should coordinate the development of a national strategy for the water sector based on the integrated river basin management principle. Such a strategy should also be agreed upon by other stakeholders.

Transboundary initiatives are encouraged in order to pave the way for international cooperation especially within the Kura river basin.

Azerbaijan is extremely poor in terms of water resources (low water availability per capita) and its water losses are significant, due largely to poorly constructed systems that date to before independence. Payment for water is based on nominal consumption rather than actual (metered) consumption, although a new programme to install water meters has recently been initiated. Without meters, the system is not able to reward those households or industries that would wish to save water. Water losses in distribution networks due to poor construction and the lack of maintenance are another big problem. The former system focused on constructing new facilities rather than on keeping existing ones operational. Money alone will not solve this problem; there is also a need to foster a maintenance culture.

Leaking water-distribution systems combined with intermittent supply cause polluted groundwater to infiltrate the pipes during periods of no pressure. Repairing the leaks will therefore also improve water quality. (See also discussion of irrigation in Chapter 10, land use, agriculture and desertification.)

Recommendation 7.2:
The Ministry of Ecology and Natural Resources, the Committee of Amelioration and Water Management, the water utilities and the water users should give high priority to reducing the high water losses in water-supply and irrigation systems. For this purpose, they should carry out a detailed analysis and prepare a step-by-step plan that prioritizes the work that needs to be carried out. The plan should include the following:
- *The water utilities should install water meters so that they can charge for their services on the basis of actual consumption;*
- *The Ministry and the water utilities should launch awareness campaigns to encourage water conservation in home installations and industrial enterprises;*
- *The water utilities should repair leaky pipes in the water-supply networks; and*
- *The State Committee of Amelioration and Water Management should reconstruct the irrigation infrastructure.*

Rivers, groundwater and the Caspian Sea are severely affected by the discharge of untreated or poorly treated waste water. This is a threat to human health and to the environment. Waste water is collected from 72% of the inhabitants of Baku but less than 50% of it is treated. Sumgayit's treatment plant is on the brink of collapse, and none of the plants in the secondary towns is working. Out of 40 operating industries in Baku and Sumgayit 33 have insufficient treatment (mechanical only or no treatment). Industrial water pollution has gone down since the political change but only due to the collapse of many industries. If these industries start up again, the facilities will be in worse condition than before and in need of major rehabilitation or reconstruction.

Recommendation 7.3:
The Ministry of Ecology and Natural Resources should ensure that the amount of untreated or poorly treated domestic and industrial waste water is reduced. To this end,
(a) The Ministry, in cooperation with the executive powers, should carry out an analysis and prepare a step-by-step plan with clear priorities;
(b) The respective executive powers should rehabilitate their sewage systems and waste-water treatment plants and/or build a new one; and
(c) Industries should be required to pretreat their waste water properly before discharging it into municipal systems.

Although most of the legal framework was updated after independence, a number of regulations and norms from the previous system still apply. Some of these are inexpedient in a system where resources, e.g. energy, are charged at cost. The SNIP norms lead to an excessive use of resources: the per capita consumption rates are at least 100% higher than western standards and so are the system requirements for water storage and transmission capacity. The present norms for waste-water treatment are unrealistic and much higher than international standards, i.e. Azerbaijan requires maximum 6 mg BOD/l compared to the EU standards of 25 mg/l.

Charges for the abstraction of water have not been adjusted since 1993. Due to the high inflation in the mid-1990s, the charges have lost their real value and the money is no longer collected. The intentions behind the system of promoting the efficient use of resources and at the same time financing water management and monitoring activities are thus not fulfilled.

Recommendation 7.4:
(a) The Ministry of Ecology and Natural Resources should review and adjust the system of norms and standards. SNIP norms should be replaced by international norms that will lead to more feasible solutions. Waste-water discharge regulations should be harmonized with international, e.g. EU, standards.
(b) Water-user charges should be increased to account for inflation.

Figure 7.1: Main rivers in Azerbaijan

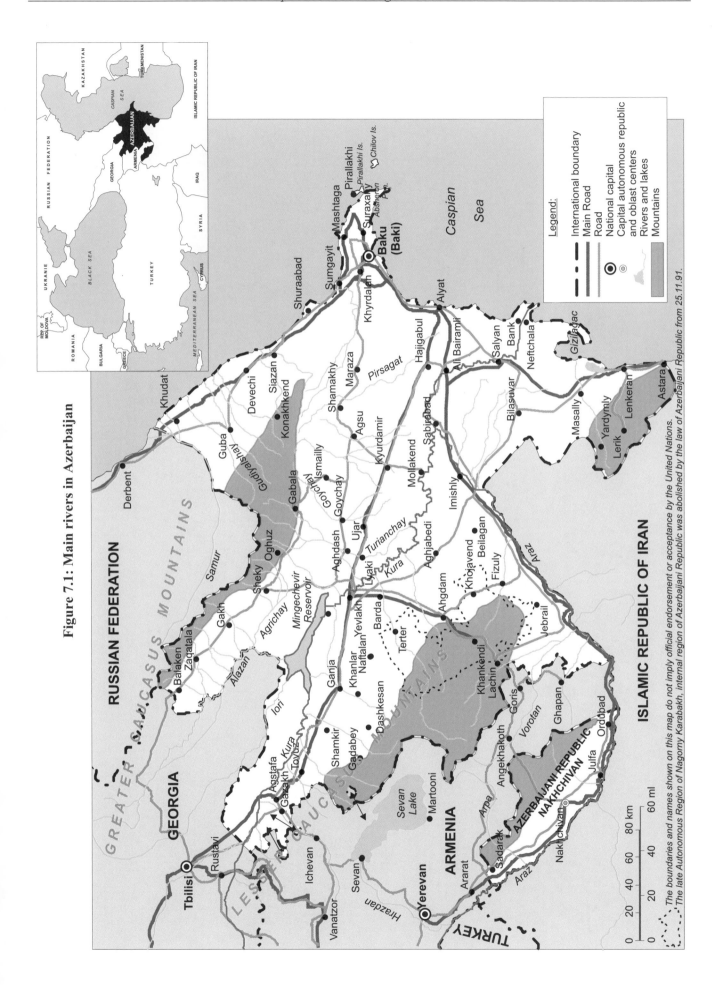

Chapter 8

SELECTED CASPIAN SEA ISSUES

8.1 Introduction

Geography

The Caspian is the largest inland body of water in the world, containing some 44% of the globe's inland waters. Physically, the Caspian Sea is one of extremes. Its salinity varies greatly. Air temperatures are likewise extreme, from summer highs in the mid–30° C range to minus 20° in the north Caspian, causing icing each winter. Morphologically, the Sea is divided into three parts, the northern shallow part (average depth 6 m), the middle section with an average depth of 190 m (maximum depth 788 m) and the southern Caspian with a maximum depth of 1025 m. Although the basin is well oxygenated in general, the vast variations in river flow may contribute to periods of deep-water anoxia from time to time.

The Caspian Sea occupies a deep depression on the boundary between Asia and Europe with a water level at present 27 m below sea level. It is approximately 1200 km long with a maximum breadth of 466 km, contains 79,000 km³ of water, and has a total coastline of more than 7000 km. The Caspian is fed by five major rivers or river groups: in the north the Volga (80% of total inflow) and the Ural (5%); in the west the Terek, Sulak and Samur (4-5%) and the Kura (7-8%); and, in the south, the short mountain rivers from the Iranian Alborz range (4-5%).

Azerbaijan has more than 800 km of coastline along the Caspian Sea and almost the entire country is part of the Sea's catchment area.

The littoral States of the Caspian Sea are the Republic of Azerbaijan, the Islamic Republic of Iran, the Republic of Kazakhstan, the Russian Federation and Turkmenistan.

8.2 Main Environmental Issues

Azerbaijan's priority concerns for its Caspian region were identified in its National Caspian Action Plan (2002). The Plan was developed by a national working group under the Ministry of Ecology and Natural Resources with the assistance of the Caspian Environment Programme. Its priority concerns include:

- The reduction in some commercial fish stocks, including sturgeon;
- The degradation of the coastal landscape and the destruction of coastal habitats;
- The threat to biodiversity;
- The general deterioration of environmental quality;
- Invasive species; and
- Contamination due to oil and gas production as well as other industrial, agricultural and municipal sources.

Water levels

The Caspian Sea is subject to considerable water level fluctuations. A decrease of 3 m was recorded between 1930 and 1977 and even much more in geological times. Between 1978 and 1996 an increase of 2.5 m was recorded, but, after 1996, levels fell by 0.5 m. These water fluctuations are mainly caused by natural phenomena but are also subject to anthropogenic effects. Evaporation is generally high, in the order of 1 metre per year, and is one of the contributing factors to water level fluctuations. However, the primary contributing factor is fluctuations in the flow from the Volga, which are tied to climatic changes. The water level fluctuations have a significant adverse impact on coastal biodiversity and infrastructure, particularly in the shallow, northern part, where loss of land and flooding are of real concern.

Predictions indicate that the levels will remain relatively unchanged over the coming years. This would justify continued monitoring and long-term planning.

Biodiversity

The Caspian Sea's biodiversity is low across all phyla compared to other seas, but, due to its historic isolation, endemism is high. Approximately 40% of the species found in the Caspian are endemic and the potential loss of global biodiversity is high. The history of Caspian flora and fauna is one of isolation and introduction. The influx of two major species, Sturgeon and Caspian seal, occurred during temporary linkages with the Mediterranean Sea through the Black Sea, and with the Arctic in past geological times. Subsequent isolation of the Caspian affected these two sources of biological populations, creating the high endemism and speciation characteristic of the system. Quantifiable data on the status of the biodiversity of the Caspian Sea are scarce. In recent years no systematic monitoring of biodiversity, except for research of fish stock and invasive species in connection with fisheries productivity, has been undertaken by the Caspian States; even population numbers of flagship species such as the Beluga sturgeon and the Caspian seal are in dispute. This lack of knowledge is in itself a major problem. Other threats include habitat erosion and degradation – again observed but not measured – habitat fragmentation, unsustainable use of key species, pollution and invasive species.

The decline in Caspian biodiversity is most visible in the loss and reduction in the number of the hallmark species and decline in habitats, such as the Caspian seal. High concentrations of certain organic chlorinated pollutants were also found in the tissues of many dead specimen. The current health status of the population, estimated to have been more than 1 million in the early part of the previous century and today numbering between 30,000 and 400,000 depending on whose estimates are to be believed, is uncertain. The sturgeons, in particular the giant Beluga, are threatened, due to overfishing and the loss of spawning grounds resulting largely from dam construction on the major rivers of the Caspian.

The Turan tiger is one of eight tiger ecotypes known in the world. However, this species is thought to have become extinct in recent times. The demise of this major hallmark species reflects the massive loss of habitat in the region, from the former vast reeds of the Iranian coastline and loss of wetlands in Kazakhstan and Turkmenistan, to the urbanization of the coast of Azerbaijan.

In the past 50 or more years, many species have been introduced into the Caspian and associated habitats, some intentional, others accidental. They have also had an impact on regional biodiversity.

Species have invaded via water transport routes. One major example of accidental introduction is the *Mnemiopsis leidyi* documented for the past few years in the Caspian Sea. It is potentially the most damaging and most acute threat so far recorded and may have already irrevocably changed the composition of the zooplankton of the Caspian. The ctenophore *Mnemiopsis leidyi* is a comb jelly which originates from the brackish waters off the southwest coast of America and which it is believed was transported first to the Black Sea and now to the Caspian in the ballast waters of seagoing vessels. The *Mnemiopsis* invasion of the Black Sea devastated the fragile fisheries ten years ago and it now threatens the Caspian Sea. The commercial fishing industry fears for the loss of kilka and other valuable fisheries, with consequent effects on livelihoods, food for the local population, and food sources for the Caspian seal and the sturgeon populations. Observations that the growth of *Mnemiopsis leidyi* biomass in the Caspian Sea is even faster than in the Black Sea support the need for rapid action. *Mnemiopsis leidyi* has already reached large densities in some southern parts of the Sea, surpassing those recorded in the Black Sea at its peak. Managing invasive species is critical to the overall health of the Caspian's biodiversity.

Fishing

The Caspian region is internationally renowned for its fisheries, and specifically for the delicacy of its caviar. The Caspian fisheries also provide much needed protein to the diets of the coastal residents. The consumption of sardines and kilka as well as sturgeon and sturgeon by-products is important to the region. The Caspian creates thousands of jobs for the fishing industry. The recent decline in fisheries has increased tensions among regionwide stakeholder groups. While steps are being taken by the Ministry of Ecology and Natural Resources to defuse these potential tensions, further efforts must be made. A number of activities are planned to be implemented by the Caspian Environment Programme.

Fishing has different levels of importance for the littoral countries. It makes up only a small part of Azerbaijan's formal economy, although the informal fishing industry is believed to be substantial. The total catch in the southern Caspian Sea over the past two decades has increased due to the rapid growth of the herring fishery, while the sturgeon harvests have slumped.

Uncontrolled poaching has devastated the sturgeon populations that were already in decline for a variety of reasons. Other fish are also under threat.

Pollution

Since the Caspian is an enclosed body of water, it has limited carrying capacity compared to other bodies of water. Pollution entering the Caspian is either biogeochemically altered, or remains in the Sea for years; none escapes and dilution is limited from external buffering waters. The circulation of the Caspian Sea is typical of enclosed seas and consists of a number of quasi-permanent cyclonic (anticlockwise) gyres that transport water and materials across boundaries, as well as smaller anticyclonic and cyclonic gyres that come and go depending on the winds, water inflows, buoyancy fluxes and other driving forces. Thus, pollution entering the Caspian Sea from the Volga and other major rivers, for instance, is ultimately distributed through the Sea and its sediments.

In the former Soviet Union, water and sediment quality measurements were taken on a regular basis and with good coverage; however, after the break-up of the Soviet Union, monitoring became fragmented and irregular as the countries in transition in the past decade have shifted budget resources away from the State Hydrometeorological Service that performed these measures. Hence the database has declined and the Soviet era findings are somewhat suspect. In Azerbaijan, the situation is changing, and the number of monitoring surveys has increased. Over the same period the flux of pollutants into the Caspian has changed, with a drastic reduction in industrial and agricultural activity in Turkmenistan, Kazakhstan, the Russian Federation and Azerbaijan. Little historical data are available on offshore waters and bottom sediments. The advent of international oil and gas exploration and accompanying environmental baseline and impact studies,

as well as various international projects and donor programmes have provided some new data during the past decade that permit some analysis.

A sediment quality analysis under the auspices of the Caspian Environment Programme showed high natural concentrations of certain heavy metals in sediments (owing to geological sources, e.g. nickel, cobalt and other metals), and persistent organic pollutants (e.g. DDT and its breakdown products, some other pesticides and insecticides). In a large area from Baku and south past the border, mercury levels in the sediment are above the effects range-low (ERL) 0.15 µg/g and DDT in the sediments along most of the coastline is above ERL (1600 pg/g).

With the exception of hot spots, no significant accumulations of petroleum hydrocarbons (PAHs) were observed in the coastal waters, nor were PCBs noted at a concentration posing human health or ecological risk.

A review of those reliable data that do exist, including data from sediment and ecotoxicological surveys undertaken as part of the Caspian Environment Programme, do not indicate a highly stressed environment, but of course there are hot spots. Nutrient loading does not appear to be a regional problem, although on the Iranian coast eutrophication is observed. Some heavy metals (aluminium, cadmium, chromium, nickel, copper and arsenic) are found at comparatively high levels throughout the Caspian sediments, but this distribution suggests the source is due to the regional geology rather than pollution. Elevated levels of mercury (in the range of 0.14-0.45 µg/g), lead and chromium indicate local pollution sources superimposed over the regional signature. Levels of agrochemicals, in particular DDT and endosulphans (organochlorine pesticides), are a major cause for concern in the Caspian. Although a banned substance, DDT and its breakdown products have been detected at high levels in sediment analyses by the Caspian Environment Programme, indicating continued use of the chemical. DDT was also detected at relatively high levels in the tissues of seal and fish in autopsies undertaken by the Programme.

Indirect evidence indicates that pollution for instance from oil slicks causes widespread hydrocarbon contamination. Furthermore, the

Ecotox Project recently reported that all fish sampled from Azerbaijan tested positive for cytochrome P-450C activity, indicating that the fish had been exposed to PAHs at some time in the recent past. However, the sources of these hydrocarbons are not clear: they may come from active oil and gas operations, from flooded contaminated area on land (due to inundation by higher water levels and surges), marine transport, even from natural seeps (the mud volcanoes emit hydrocarbons to the environment along the Absheron ridge separating the south Caspian from the middle Caspian), rivers, and perhaps other industry.

In summary, pollution threats may include: contaminants sequestered in the major impoundments on the Volga, above Volgograd, and the Kura; continued and increased use of banned agrochemicals; potential widespread hydrocarbon pollution, with the anticipated expansion of oil and gas development; and acute damage from oil and other hazardous spills. In some Caspian littoral States, pollution load data are very poor, and there is a general lack of knowledge regarding the specific issues in basin hot spots.

8.3   Policy objectives and management

Policy framework

Caspian Environment Programme

In 1994, the five littoral States of the Caspian Sea (Azerbaijan, Islamic Republic of Iran, Kazakhstan, Russian Federation and Turkmenistan) adopted the Almaty Declaration on Cooperation for the Environmental Protection of the Caspian Sea Region, in which they recognized that the Caspian Sea region was facing severe environmental problems with serious social and economic impacts. They decided to undertake coordinated action and called for the assistance of the international community.

As a response, a joint UNDP/UNEP/World Bank Caspian initiative was taken in 1995. Its purpose was to coordinate the activities of international organizations for the environmental protection of the Caspian Sea and resulted in May 1998 in the official launching of the Caspian Environment Programme.

The Caspian Environment Programme is a comprehensive long term strategy for the protection and sustainable use of the Caspian Environment. The first phase of CEP was conducted from 1998 to 2002 with the following main goals:

(a) Development of a regional coordination mechanism to achieve sustainable development and management of the Caspian environment;
(b) Completion of a Transboundary Diagnostic Analysis (TDA) of priority environmental issues;
(c) Formulation and endorsement of the Strategic Action Programme (SAP);
(d) Adoption of National Caspian Action Plans.(NCAPs)

The TDA, the NCAPs and the SAP are the result of extensive regional consultation process based on essential inputs from the five Caspian littoral states and the international partners. The process has included a causal chain analysis, stakeholder analysis and gap analysis. The five NCAPs and the regional TDA are the pillars of the SAP and they should be considered together as a whole.

The TDA identified eight major perceived environmental problems and issues. These were later refined through regional consultation into four priority environmental regional concern areas further elaborated in the SAP as regionally agreed Environmental Quality Objectives (EQO) in a transboundary context. The four EQO correspond to the four main areas of concern, whilst one additional EQO has been identified for the cross-sectoral issue of strengthening the involvement of all stakeholders. The overall five EQO spelled out in the SAP are the following:

• conservation and sustainable use of bioresources;
• conservation of biodiversity;
• improved water quality for the Caspian Sea;
• sustainable development of the coastal areas; and
• strengthen stakeholder participation in Caspian environmental stewardship.

The SAP is a regional policy framework laying down principles of regional environmental management and cooperation. It identifies national and regional interventions to address the priority environmental concern areas as well as the necessary financial resource and institutional structure required for the implementation of the priority actions for the next 10 years (two separate 5-years periods).

Implementation of SAP is the responsibility of the Caspian States independently as component of their NCAP, and collectively as part of the CEP. The CEP Steering Committee has the responsibility of monitoring and reporting on SAP implementation progress.

In November 2003, the GEF council approved the second phase of GEF support to the CEP. The main four objectives of the second phase of CEP include:

(a) To initiate the implementation of the SAP in three priority areas: biodiversity, invasive species, and persistent toxic substances;
(b) To continue with specific capacity building measures to ensure a regionally owned CEP coordination mechanism capable of full implementation of SAP and regional coordination of the NCAPs;
(c) To strengthen environmental legal and policy frameworks operation at the regional and national levels;
(d) To achieve tangible environmental improvements in priority areas by implementation of small-scale investments.

The second phase is expected to facilitate the continued transition to enhanced governance and execution by the littoral states themselves, with the international partners playing a supportive role. The full ownership of the CEP by the Caspian littoral States is the ultimate aspiration of the project.

The European Union is also in the process of identifying further support to the CEP and two separate projects focusing on Coastal Sustainable Management and Fisheries are in the pipeline.

The Strategic Action Programme will also be the basis for the implementation of the future protocols to the Framework Convention for the Protection of the Marine Environment of the Caspian Sea, which was adopted by all five littoral States and opened for signature, in Tehran, on 4 November 2003.

The Caspian Framework Convention aims to reduce pollution and includes provisions for the control of pollution from land-based sources, seabed activities, vessels, dumping and other human activities. General provisions for coastal zone management, sea level fluctuation and environmental emergencies are also contained in it.

The Caspian Framework Convention establishes the general principles and the institutional mechanism for environmental protection cooperation in the region. It is expected to serve as the overarching legal instrument for regional environmental cooperation, while the real "teeth" will be provided for in subsequent protocols. Currently, it foresees the development of at least seven protocols for the following specific concerns:

- Pollution from land-based sources;
- Pollution from seabed activities;
- Pollution from vessels;
- Pollution caused by dumping;
- Protection, preservation and restoration of marine living resources;
- Sea level fluctuations;
- Environmental impact assessment.

Given its general framework character, it is foreseen that the convention can accommodate any additional specific protocols that its Parties might consider necessary.

The adoption of the convention has paved the way for a continuation of the Caspian Environment Programme, which has the following main objectives:

- Ensure the sustainable development of economic activities;
- Decrease pollution levels and improve the quality of the environment and bio-resources of the Caspian region;
- Improve and rehabilitate the ecosystems of the Caspian Sea and conserve its biodiversity;
- Ensure environmental safety in the region and conserve an environmental quality compatible with sustainable human development.

The Caspian Environment Programme has been very successful in providing a comprehensive analysis of the state of the Caspian Sea. It appears that cooperation between it and Azerbaijan has been very good, contributing to the success of the Programme.

Legal framework

The laws relevant to the Caspian Sea are generic for the environment in Azerbaijan and are covered in chapter 1 and elsewhere, as appropriate.

An environmental impact assessment (EIA) should precede any activity that may have an impact on the environment. Azerbaijan was one of the first Caspian littoral States to ratify the EIA Convention in a transboundary context, and it has gained substantial experience in this area. The EIA regime appears to function well for major activities in the Caspian Sea. The EIAs, however, are isolated events and overall sectoral policies are not scrutinized for their environmental impact through strategic environmental assessments.

The establishment of a Caspian environmental fund has been considered to secure sustainable interventions for the Caspian Sea, but the issue has not been pursued. The Strategic Action Programme has estimated the national cost of addressing the actions that it recommends for each of the countries. The establishment of a fund to implement those recommendations could help to mobilize adequate external funding.

Several multilateral environmental agreements provide useful legal frameworks for national legislation relevant to the protection of the Caspian Sea. Azerbaijan is ready to accede to the 1973 International Convention for the Prevention of Pollution from Ships (MARPOL). The Convention on Biological Diversity will be a useful tool to address the threatened biodiversity in the Caspian Sea. All three protocols adopted and signed at the Kiev Ministerial Conference in May 2003 are relevant to the environmental management of the Caspian Sea.

Institutional framework

The development and implementation of State environmental management policy for the Caspian Sea to study, reproduce, use and protect natural resources and provide environmental safety is the responsibility of the Ministry of Ecology and Natural Resources.

Nearly 10% of its budget (around US$ 1 million) is assigned annually for monitoring, control and research in connection with the Caspian Sea.

The policy of the Government focuses on achieving consensus on the legal status of the Caspian Sea, regulation of fisheries, prompt verification of agreement on the use and conservation of the Sea's bio-resources,

development of common environmental standards for the use of its resources, emergency and oil spill response and preparedness, and extension of a zero-discharge principle to the entire Caspian.

However, monitoring and reporting appear to be historically determined rather than a dynamic response to the current situation in the Caspian Sea. Given the findings of the Caspian Environment Programme, a sufficiently comprehensive monitoring system will be needed as a basis for policy decisions and the reported data should be made widely available to the public.

One of the most important directions of the policy is to improve the effectiveness of the environmental management system and to create conditions for sustainable development. Significant changes were already made in 2001, when a presidential decree (23 May 2001) abolished the State committees of ecology, geology, hydrometeorology, and Azerbmeshe Industrial Association. As a result of this restructuring process, the Ministry of Ecology and Natural Resources was established. Later, when the Azerbalig State Corporation was terminated, its functions of regulations and reproduction of fish stock were transferred to the MoENP.

Azerbaijan, in the first phase of the Caspian Environment Programme, hosts the Programme's regional thematic centres on pollution control and database management. The Programme Coordination Unit was transferred from its original location in Baku to Tehran at the end of December 2003 as originally foreseen in the first phase of the Programme.

Each littoral State will chair and host one of the advisory groups; exactly which one will be decided during the next phase of the Programme.

8.4 Conclusions and recommendations

The sound environmental management of the Caspian Sea is a high priority for Azerbaijan. The legal and institutional framework covers the Caspian Sea, and the Government is finalizing its National Caspian Action Plan as the major policy document for the management of the lake's environmental resources.

The Caspian Environment Programme's Strategic Action Programme outlines what the littoral States of the Caspian Sea have to do to protect and manage it in an environmentally sustainable manner. In the future, the selection and development of protocols to the framework convention will be harmonized with the requirements of the national Caspian action plans, the Strategic Action Programme and the framework convention. Azerbaijan is now working on making its National Caspian Action Plan fully consistent with the Caspian Environment Programme's Strategic Action Programme.

Recommendation 8.1:
The Ministry of Ecology and Natural Resources should facilitate the process of adoption of the National Caspian Action Plan and ensure its consistency with the Caspian Environment Programme's Strategic Action Programme. It should also support the Strategic Action Programme at the regional level.

The political process leading to the Framework Convention for the Protection of the Marine Environment of the Caspian Sea has been long and difficult. Adequate protection and management of the Caspian Sea – and initial funding to do so – hinges upon timely ratification of the convention.

The second phase of the Caspian Environment Programme will be approved only if Governments are ready to accept ownership of the convention and the Strategic Action Programme, and demonstrate it by providing adequate financial and institutional support.

Recommendation 8.2:
The Ministry of Ecology and Natural Resources should actively pursue and solicit support from the other Caspian States to ratify the Framework Convention.

In order to secure international funding for future activities the idea of establishing an environmental fund has been put forward by the Caspian Environment Programme but not yet thoroughly discussed and assessed by the Caspian littoral States. The idea should be further analysed to identify the potential sources of funding and issues of governance of such a fund, leading to the establishment of such a national mechanism.

The establishment of a fund would also be a clear demonstration of national commitment to solve the environmental problems of the Caspian Sea.

Recommendation 8.3:
The Government of Azerbaijan should cooperate with the other Caspian littoral States in establishing an environment fund for the Caspian Sea, specifying potential sources of financing and institutional responsibility.

MARPOL is the main international convention covering the prevention of pollution of the marine environment by ships from operational or accidental causes. It is a combination of two treaties adopted in 1973 and 1978 respectively and updated by amendments through the years. Azerbaijan has not yet ratified MARPOL, but this Convention could provide an important and useful legal framework for national legislation relevant to the protection of the Caspian Sea, and its implementation would facilitate Azerbaijan's efforts to reach the policy goals elaborated within the National Caspian Action Plan.

Azerbaijan did ratify the Convention on Biological Diversity in 2000, and it is important that this Convention should be fully implemented, including in the context of the National Caspian Action Plan.

Recommendation 8.4:
(a) The Ministry of Ecology and Natural Resources should promote the ratification by Azerbaijan of MARPOL as soon as possible;
(b) The Ministry of Ecology and Natural Resources should ensure full implementation of the Convention on Biological Diversity, including in relation to the biodiversity of the Caspian Sea.

The environmental monitoring system should pay greater attention to monitoring of the state of the Caspian Sea, including pollution levels and biodiversity, to provide a basis for sound policy decisions. The results of such monitoring should be made widely available to the public. (See chapter 3, on environmental information and public participation.)

Recommendation 8.5:
The Ministry of Ecology and Natural Resources, when finalizing its State programme for strengthening environmental monitoring, should include a plan for strengthening the monitoring of pollution levels of the Caspian Sea and for disseminating its results.

Figure 8.1: Topography of the Caspian Sea region

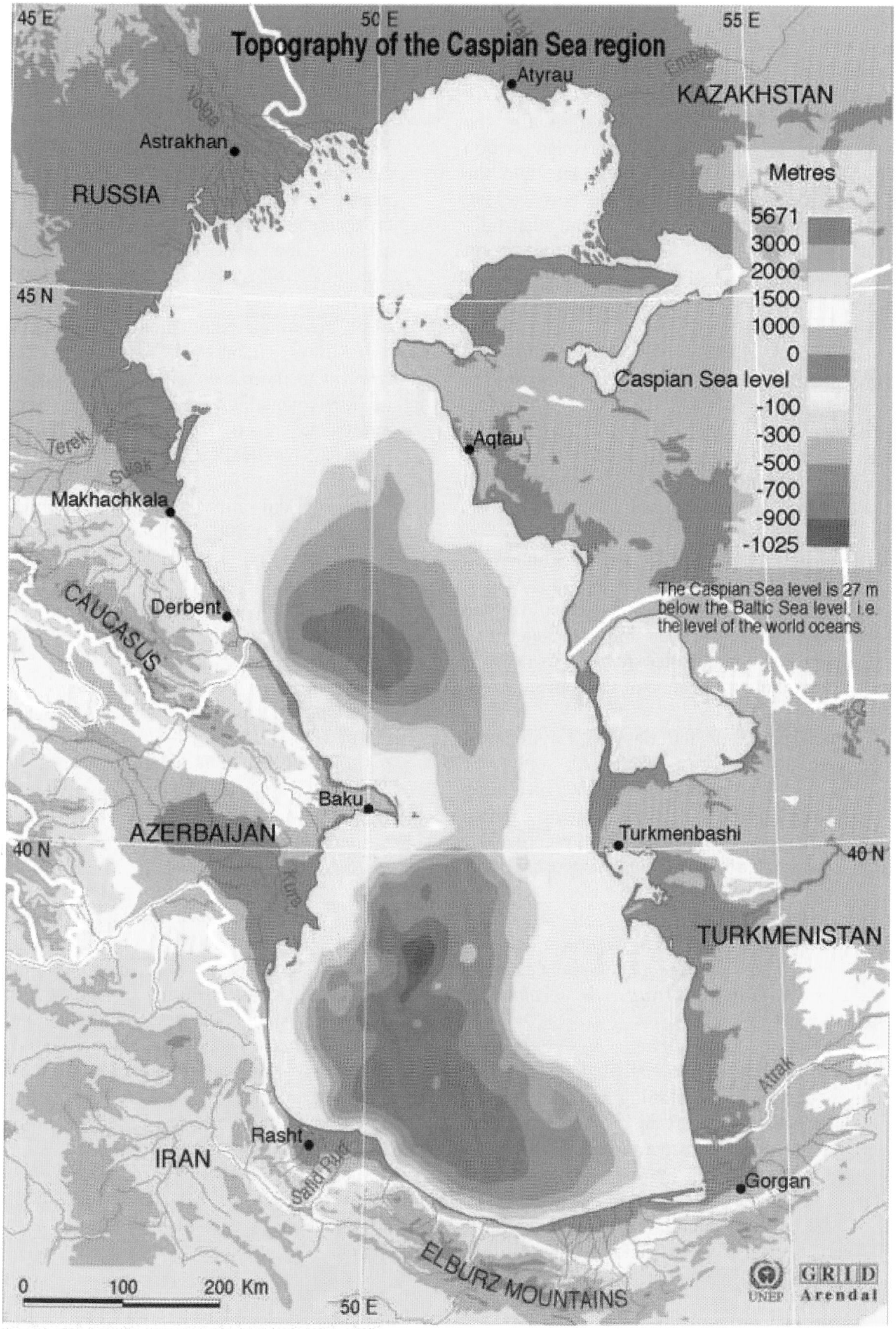

Source: The Caspian Environment Programme, 2002.

Figure 8.2: Flooding with continued water level rise

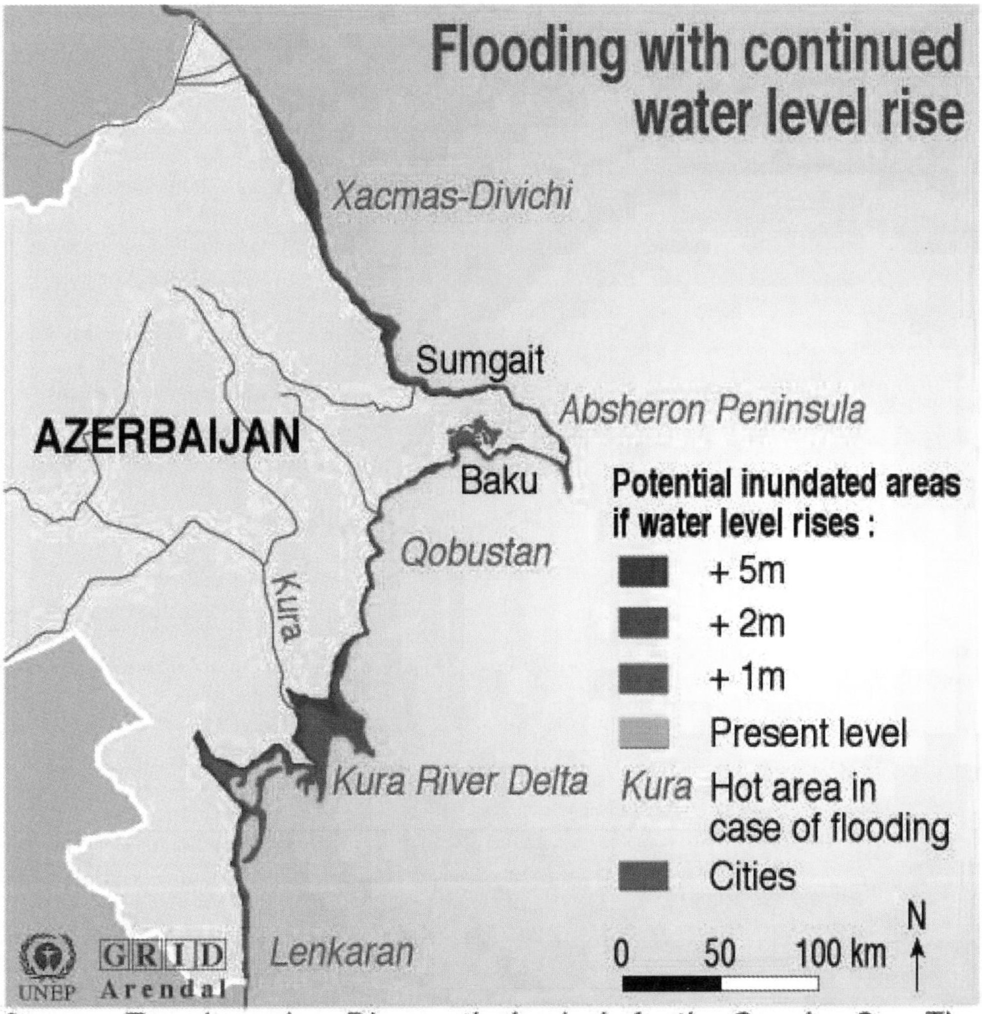

Source: *Transboundary Diagnostic Analysis for the Caspian Sea*, The Caspian Environment Programme, 2002.

Table 8.1: Targets and expected results of EQOs from the NCAP

Expected results of environmental objectives	Targets
1. Sustainable economic use of the Caspian and coastal area The implementation of new technologies and equipment will allow to continue development of the offshore oil fields with minimal risk and transport the product in a safest way. Development and implementation of measures to reduce environmental pressure on sensitive areas, allocate economic facilities, reduce pollution of surface, ground waters and sea water	1- development and implementation of a plan of integrated coastal management by 2012 2- provide safety of exploration, production and transportation of hydrocarbons by means of the advanced technologies and modern equipment that allows to prevent pollution of biosphere by 2030 3- reduce impact of agriculture on the surface and ground waters by 2007 by means of new agricultural technologies, establishment of new norms for cattle grazing. 4- reduce pollution of the sea by domestic and industrial sewage in Baku and Sumgayit by 2012
II- Sustainability of the environment of the Caspian including conservation of biodiversity Implementation of the project will increase the area of protected zones by a factor of 2,5; rehabilitation of spawning grounds of sturgeon and other valuable commercials species will provide commercial return of about 8 thousand ton; establishment of biological centers for breeding of rare and endangered species of fauna and forestation will create favorable conditions for sustainable use of resources of the Caspian and coastal area, along with the conservation of biodiversity of ecosystems	1- establishment of a national park Shah Dag with affiliate in Yalama by 2007 2- rehabilitation of natural spawning grounds in Kura and Araz Rivers on the area exceeding 150 hectares by 2012 3- establishment of a biological center for reproduction of rare and endangered species with five affiliates in different areas by 2007 4- prevent degradation of coastal area, fix sands and eroded slopes by 2005 5- increase public awareness about value of biodiversity of the Caspian sea by 2005 6- development of guidelines for economic assessment and mechanisms of compensation of biodiversity loss by 2005 7. Increase capacity of sturgeon hatcheries up to 30 million per year by 2012
III- High quality of sea water, ground waters and surface waters The implementation of the projects will create favorable conditions for fast reduction of pollution of sea water caused by domestic and the industrial discharges, increase recreation capacity of the coastal zone, rural population will use high quality drinking water. Cleaning of sea water will increase recreation potential	1- quality of drinking water will improve in rural areas by 2012 2- regulation of industrial, municipal and agricultural discharge into the environment by 2007 3- reduced areas with polluted surface and ground waters of Absheron peninsula by 2012 4- higher quality of coastal waters of recreational zones in Yalama, Absheron and Lenkaran 2012 5- capacity building for emergency response and oil and chemical spill response onshore and offshore by 2007
IV- Sustainable combined use of the coastal area of the Caspian Improvement of hydrological and biological regime of reserves and valuable wetlands will create conditions for the rehabilitation of wintering sites of migrating bird pieces, allow conservation of rare or endangered species	1- introduction of feasible norms of recreation pressure on landscape complexes by 2007 2- improved hydrology for Sheehan and provision of sustainable use of wetlands of Kyzyl Agach reserve by 2007 3- establishment of Taurus complexes for 5 000 people (Yalama, Apsheron, Lenkaran) by 2012 4- opening of a Center Of Environmental Training in Baku with affiliate in Kuba by 2005
V- Higher life standard Implementation of the planned actions is connected with general economic raise for the population of coastal area, new job opportunities, higher life standards, improved health conditions and wider range of medical service and use of medicines. Construction of tourist complexes and development of ecotourism will create favorable conditions for recreation, as well as stimulate development of agriculture, food and processing industries	1- rehabilitation of production potential and capacity of processing of agricultural products in coastal area by 2007 2- development of mechanism and creation of favorable conditions to attract investments into the local sectors of economy by 2005 3- introduction of measures for sustainable use of fisheries by 2012 4- organization of yearly sales in border area of coastal zone, improvement of trade exchange in the Caspian region by 2005 5- presentation project on ecotourism on the example of national park in Shah Dag by 2007 6- assistance to implementation of national health program by 2012 7- expansion of environmental education by 2005 8. Access to information and public participation in the development of NCAP

Chapter 9

BIODIVERSITY AND FOREST MANAGEMENT

9.1 Introduction

Azerbaijan occupies the territory of the southeastern Greater Caucasus, the eastern Lesser Caucasus, the central hills and lowlands with the Kura and Araz valleys, and the Talysh Mountains in the southeast. The region is internationally recognized as one of the 25 most important global hot spots due to its high biological diversity (especially endemic, rare and relict species) and very endangered ecosystems.

Since 1992, Azerbaijan has made significant progress in international cooperation on nature protection by becoming Party to five international conventions dealing with biodiversity conservation (see *International agreements* below). It brought into force general laws on environmental protection (including nature conservation), wildlife and protected areas. These laws need further harmonization with international standards, in terms of better elaboration of implementation mechanisms.

The new Ministry of Ecology and Natural Resources is faced with the difficult task of designing a nature conservation policy to improve biodiversity management under economic conditions that have led to the overexploitation of land resources.

9.2 Species and habitats

The geographical position of Azerbaijan, at the juncture of Europe and Asia, caused the development of very diversified terrestrial ecosystems in which species of both continents can be found (southern species such as Persian gazelle, leopard, and striped hyena, and typically North European species such as wolf, brown bear, chamois, lynx and wild boar) (see tables 9.1 and 9.2). It also provides habitats to some endangered bird species of Southern Europe, such as the white-headed duck, ferruginous duck, and

marbled teal, and to some globally endangered ones (lesser white-fronted goose). This, in particular, increases the importance of its wetlands, which serve as wintering and nesting areas. Endemic sea species, sturgeon and other fish species, as well as seal, characterize the biodiversity of the Caspian Sea.

The Red Book of Azerbaijan was published in the 1980s and is outdated. The Ministry of Ecology and Natural Resources started updating it in 2003. Completion depends on financing and inventory dynamics. Support is expected from the World Bank and the United Nations Development Programme (UNDP).

There are six ecological regions in Azerbaijan: the Greater Caucasus Mountains, the Lesser Caucasus Mountains, the Kura-Araz valley and floodplain, the Talish-Lenkeran mountainous zone, Absheron peninsula, the Caspian Sea coastal lowland zone and the area encompassing the Nakhchivan Autonomous Republic. The Greater and Lesser Caucasus are discussed together as they are very similar.

Table 9.1: Number of species by taxonomy groups

Flora:	Number of species
Flora	
Endemic	270
Medicinal plants	800
Fauna:	
Insects	14,000
Arachnids	11,000
Vertebrate (of which there are more than 100 fish species)	600
Amphibians	9
Reptiles	54
Birds	360
Mammals	102

Source: Nature Conservation in Azerbaijan Republic, document prepared in 2000 for the Council of Europe by the State Committee of Azerbaijan Republic on Ecology and Nature (follow up of the Convention on the conservation of European wildlife and natural habitats)

Table 9.2: Number of endangered animal species

Group	ARDB	RDB / FSU	IUCN	Total
Mammals	14	10	25	32
Birds	37	24	14	53
Reptiles	8	6	6	15
Amphibians	5	2	1	6
Fishes	5	n.d.	19	19
Insects	40	n.d.	11	49

Source: Azerbaijan Red Book (ARB), Red Book of the Former USSR (RB/FSU) and IUCN Red Books (1996).
Note: n.d. = no data

The Caucasus Mountains (Greater and Lesser) stretch at altitude of 300 to 2500 m. These are dominated by oak mixed with, for instance, hornbeam, lime, sweet chestnut, ash, and others, at lower altitudes. Oak, birch, beech and maple dominate in the higher zones. At altitude of 2500 to 3000 meters grass meadows and rhododendron shrub, as well as rocky plants can be found.

Mountain steppes cover lower altitudes, and are used for agriculture (crops, vegetables, fruit trees and fodder plants). There are many species that are remote ancestors of domestic fruit trees (e.g. pear and almond trees). Dry scrub forests of Juniper, pistachio, maple, and almond are combined with scrub species.

The Kura-Araz valley and floodplain includes semi-desert vegetation (dominated by wormwood). Steppe vegetation occurs in the lowlands and foothills, dominated by grass and thorny shrubs, within which small areas of endemic Eldar pine (*Pinus eldarica*) can be found. This zone is also used for agriculture.

Riparian (*tugay*) forests occur along riverbanks. Wing nut, oak and poplar prevail. These forests are especially endangered in the Kura valley, because the water regime has been changed by the construction of a hydropower plant.

The Talish-Lenkeran zone is found along the western part of the Caspian coastline and the Talish Mountains area. This area is well known for relict forests (*Quercus castanifolia*, *Zelkova carpinifolia*, Caucasian persimmon, Girkan poplar and hornbeam).

The Caspian Sea provides 800 km of coastline in Azerbaijan. Assessments indicate that the Caspian Sea provides habitat for 90% of the world's sturgeon population, among other marine fauna and flora species. The coastal wetlands and islands are important nesting and wintering sites for waterfowl species.

9.3 Pressures on flora, fauna and their habitats

Hunting

The Scientific Coordination Council for Hunting Issues of the Ministry of Ecology and Natural Resources is responsible for permitting and controlling hunting. The Head of the Biodiversity Department chairs the Council, while other representatives are from scientific institutions, enterprises and NGOs. The Council decides on 10-year contracts on hunting grounds (30 altogether), mostly with the hunting association Azerhunters. The Zoological Institute of the Academy of Sciences, in cooperation with the Baku State University and NGOs, participates in monitoring of compliance with hunting regulations. The Ministry's Biodiversity Department calculates quotas based on these monitoring results.

Hunting is based on a ticket (permit) system, with two types of tariffs, one for foreign and the other for local hunters. Tickets can be daily, 5-day, 10-day or seasonal. Their fees are calculated by multiplying the minimum wage (5500 manats) by coefficients that vary from 3 to 55 for local hunters and from 90 to 225 plus a trophy coefficient (the highest is 730 for bear trophy) for foreign hunters. The number of animals per permit is restricted. Sixty per cent of the money collected from permits is allocated to the management of the hunting grounds. Hunting grounds are managed by hunting associations. The largest one is Azerhunters, which distributes this money to its regional units. The Biodiversity Department does not collect records on animals killed, except those on sold permits (in 2003: 1298 daily, 36 five-day, 23 ten-day and 5 seasonal).

Table 9.3: Declines in numbers of selected mammal species, from 1976/1980 to 1991/1996

Species	1976-1980	1991-1996
Caucasus deer	2,500-2,800	1,000-1,200
Roe deer	3,500-3,800	2,800-3,000
Gazelle	3,000-3,500	2,000-2,800
Wild Goat	2,300-2,800	1,200-2,000
Mouflon	1,600-1,800*	500-700

Source: NEAP Biodiversity Working Group Report.
Note: * Data from 1981-1985.

Fines are also calculated by multiplying the minimum wage by coefficients that vary from 50 to 500. The Biodiversity Department reported that poaching was a serious problem, but annual data on penalties and fines were not available. It is not clear whether these data are not recorded or not accessible. Consequently, it is not possible to assess the real extent to which animals are hunted. Decisions on hunting are made by the Scientific Coordination Council for Hunting Issues which is represented by Ministry staff, scientists and NGOs.

Table 9.3 shows decline in numbers of selected mammal species from 1976/1980 to 1991/1996. More recent figures (2003) indicate a reversal to the trend for roe deer, gazelle and mouflon. Numbers of Caucasus deer and wild goat have continued to decline.

Grazing and agriculture

Local communities, including large numbers of refugees, try to solve their economic problems by raising cattle, over which there is no control. Too many animals per unit of area degrade pastures and forest areas, especially young trees and shrubs. Forest regeneration is thus endangered, and overgrazed pastures turn to barren soil exposed to erosion. An integrated approach to forest and rangeland management is needed.

Pollution of agricultural land decreased in the past decade due to the reduced use of fertilizers and other agrochemicals, but soil erosion and salinization are still problems because irrigation systems are poorly maintained (see chapter 10, on land use, agriculture and desertification). The annual loss of water through leakage is estimated at 2.5 to 3.0 billion m^3, causing groundwater levels to rise, which in turn leads to land salinization and swamping. Salinization mainly occurs in the Kura-Araz lowlands, where more than 387,000 ha of land is now affected. As land becomes saline, the properties of the soil change, resulting in changes in the vegetation cover and consequently in dependent fauna species too.

Azerbaijan has a rich spectrum of crop plants, in terms of number of both species and variation within individual species. Wild relatives of important crop plants are frequent. There are also surviving old landraces of domestic animals.

Land degradation has a considerably negative influence on the biodiversity of wild fauna and flora. As in other countries, the development of monoculture and the introduction of new modern plant varieties and animal races are also a continued threat to traditional crop cultivars and races of domestic animals. During the Soviet period many of the landraces were exchanged for a few introduced varieties and the landraces were mostly confined to collections outside Azerbaijan. There are local populations and genotypes, in particular of fruit trees and grapes, that are still used but not identified and brought into schemes for conservation.

Commercial fishing in the Caspian Sea

The sturgeon catch decreased from 0.25 million quintals in 1980 to 0.13 million in 1990. A further decline in sturgeon catches after 1990 has been reported. At the time of the Environmental Performance Review, 23 companies and 6 individuals had received fishing permits for 2003, and 19 tons of sturgeon were caught by June 2003, which is below the national quota.

Power plants

Construction of the hydropower plant in the Kura-Araz river catchment system (Mingchevir) changed the water regime and contributed to the degradation and loss of tugay forests. Another hydropower plant, which is currently under construction in the Kura river valley (Enikend), could add to the habitat degradation. The natural flood cycle on which these forests depended was lost. Currently, a project is being developed on Kura River Delta Rehabilitation and Protection under the support of World Bank. The main objectives of the project are cleaning the Kura river delta from silt in order to facilitate migration of sturgeon and other fish species to spawning grounds, as well as conducting amelioration works at the spawning grounds and organizing measures on protection of the ecological system of the delta.

Tugay forests are among the last original white poplar habitats, and provide nesting and wintering sites for waterfowl species. The Ministry of Ecology and Natural Resources is considering measures for the restoration of these areas, including pumping and irrigation systems that would keep changes in the water levels close to natural ones. Although it is not possible to assess the size of the impacted area without a thorough EIA, the fact is that the few remaining tugay forests are very endangered and should be preserved.

9.4   Forestry

Since 2002, forest management plans have been developed by the Forest Protection and Rehabilitation Units, which is financed by the State budget, and the plans are approved by the Forestry Development Department of the Ministry of Ecology and Natural Resources. Lack of institutional experience in both forest inventory and planning is considered a major institutional problem in dealing with forestry. Prior to independence, forest inventory and planning was centralized in Moscow and the inventory fieldwork was the responsibility of the Caucasian Forest Stocks Projection Institute in Tbilisi (Georgia). The completed forest management plan was then given to Azerbaijani forest experts and based on these documents protection and other forest work were conducted. Forestry training is available at the Azerbaijani Forestry Faculty in Ganja but it is insufficient for today's management purposes. The Faculty needs further development of study curricula, especially in inventory methods and new techniques (e.g. geographical information system (GIS) tools). Currently, new forest management plans are under development by a newly established unit of the Ministry of Ecology and Natural Resources.

According to the latest forest inventory of 1984, the total woodland area of Azerbaijan was 1,213,700 ha, or 14% of the territory. By 2001, the percentage had declined, to approximately 11% of the territory, and there is concern that the actual figure could be lower because of deforestation,

military conflict and increased demand for fuelwood in remote mountainous areas where there is a shortage of natural gas supply.

The data on average annual forest harvesting obtained from interviews at the Ministry's Forestry Development Department generally correspond to the data presented in the Preliminary Assessment of Wood Production and Marketing by World Bank experts (table 9.4). During the mission, it was reported that the average annual forest cut was about 60,000 m^3 for sanitary and maintenance purposes only. However, from 2003 forest cuttings for sanitary purposes have been suspended and only forest residues are collected. It was also roughly estimated that an additional 30,000-40,000 m^3 per year is cut illegally.

The protection and rehabilitation of forests are financed from the Forest Fund, administered by the Forestry Department. Its sources of revenue are fines imposed for violating the Forestry Code and donations. It was reported that the Fund was mostly used to cover the salaries of employees in rayon offices, and there is not enough for other reforestation expenses. According to the National Programme for the Restoration and Expansion of Forests (approved by the President in 2003), about 64,000 ha should be reforested by 2008, but only about 3000 ha will be completed in 2003, owing to insufficient financing.

9.5   Forest protection

Forests play the main role in preventing soil erosion, and provide habitat for most of the terrestrial animal species. In Azerbaijan, forests are degraded owing to two major problems. These are illegal logging and grazing, both caused by economic problems and poverty.

Table 9.4: Total number of maintenance cuttings and residues rectification in 2002

Objective	Area (ha)	Area %	Yield (m^3)	Average yield (m^3/ha)	Estimated revenues in thousand manats	Estimated revenues US $
Total	**5,806**		**64,207**		**993,110**	**216,363**
Regeneration cuttings including:	4,594	100.0	53,477	11.6	976,496	212,744
Sanitary cutting	3,900	85.0	43,605	11.2	766,211	166,930
Aid to young forests	186	4.0	523	2.8	6,894	1,501
Thinning	114	2.5	1,395	12.3	21,878	4,766
Transition cutting	395	8.5	7,954	20.1	181,513	39,545
Rectification of residues	1,212	..	10,729	8.5	16,614	3,619

Source: Ministry of Ecology and Natural Resources, 2003.

Forestry inspectors report that illegal logging is widespread, amounting to 30,000 to 40,000 m^3 annually (no statistics were available to verify these figures). Owing to the energy crisis, local people and refugees cut forests to provide their households with firewood. 1600 staff members are engaged in forest protection, of which 1100 are forest rangers. The work of forest rangers is complemented by regular spot checks performed by inspectors. Each Forest Protection and Rehabilitation Unit employs one inspector (total 34

in the country). The effectiveness of the work of inspectors is weak; they are not sufficiently equipped (there is primarily a shortage of vehicles and communication tools).

Two measures for combating forest degradation are being implemented. The first one is fencing of forest areas in which there is young growth, which needs to be preserved, to keep livestock out. The other is reforestation (see above).

Table 9.5: Strict nature reserves

Name	Area (ha)		Ecoregion	Ecosystem	Purpose of establishment
	2002	2003			
Ag-Gel	5,182		Kura-Araz	Wetlands, semi-desert	Waterfowl
Alti-Agach	4,438	4,438	Greater Caucasus	90% in broad-leaved forest	Ecosystem conservation
Basit-Chay	107	107	Lesser Caucasus	Riparian forests	A unique grove of ancient plane trees
		3,139	Nakhchivan	Bata-bat lake, forest territories	Conservation of rare, endemic flora and Bata-bat lake
Gey-Gel	7,131 (including 392 ha of Eldar pine)	7,131	Lesser Caucasus	60% forests, 40% sub alpine and alpine meadows, wetlands	Conservation of mountain forests and sub alpine area of lesser Caucasus, provision of the purity of water in the Gey Gel Lake and Eldar pine
Girkan	2,904	2,904	Lenkeran-Talysh	Humid subtropical forests	Rare, endemic and relic flora
Zakalata	23,844	23,844	Greater Caucasus	Forests, alpine meadows	Protection and study of the fauna and flora of the southern slopes of the Greater Caucasus
Ilisu	9,345	17,382	Greater Caucasus	93% is broad leaved forests	Ecosystem conservation
Ismaili	5,778	16,740	Greater Caucasus	Mountain and lowland forests	Rare and endangered species
Gara-Gel	300	300	Lesser Caucasus	High mountain lake	Conservation of the lake
Gara-Yaz	4,855	9,658	Kura Araz	Lowland tugay forest	Forest conservation
Kizil-Agach	88,400	88,400	Caspian	Wetlands	Bird (mainly waterfowl)
Pirguli	1,521	4,274	Greater Caucasus	85% are forests	Ecosystem conservation
Turian-Chay	12,630	22,488	Greater Caucasus	Semi desert and arid light forests	Ecosystem conservation
Shirvan	25,800	6,232	Caspian, Kura Araz	Semi desert meadows, wetlands	Conservation of gazelle, waterfowl and plants of the Shirvan lowlands
Total area (ha)	192,235	207,037			
Percentage of the total land area of the country	2.2	2.4			

Sources : Chemonics International/USAID's Bureau of Europe and NIS, Biodiversity Assessment, 2000. Ministry of Ecology and Natural Resources, 2003.

9.6   Protected area system

Protected area categories defined by the Law on Specially Protected Natural Areas and Objects (2000) include those that are international (some natural reserves), national (State reserves and national parks) and local (natural reserves, natural monuments, zoological parks, botanical gardens and dendrological parks, health resorts). Private landowners or long-term lessees can also establish local specially protected areas. National parks, natural parks, ecological parks and zoological parks have the status of nature protection and research institutions. Protected areas are established for protection, research, monitoring, training and tourism purposes. Banned activities are listed for each protection category.

Currently, there are 37 protected areas, of which 14 are strict nature reserves, 20 are sanctuaries and 3 are national parks (see tables 9.5, 9.6 and 9.7). Strict nature reserves correspond to the first protection category of the World Conservation Union (IUCN). According to the National Report on the Ecological Network (2001), they occupy 192,235 ha (2.2% of the country). Each strict nature reserve is managed by a director and has 20 to 25 permanent staff. Sanctuaries and other protected categories (including endemic and relict forests) occupy 379,000 ha or 4.3% of Azerbaijan's territory. They have a weaker conservation status and are managed either by strict nature reserve managers or by the regional authorities. In 2003 the area of strict nature reserves amounted to 206,937 ha, the area of national parks – 84,428 ha, and the area of sanctuaries – 273,860 ha with combined area covering 7% of the country's territory.

Table 9.6: Sanctuaries

Name	Area in 2003 (ha)	Ecoregion	Ecosystem	Purpose of establishment
Garayazi Agstafa	7,167	Kura Araz	Riparian (tugay) forests	Mammals and birds preservation and restoration
Barda	7,500	Kura Araz	Riparian forests and lowlands	Game birds
Sheky	10,350	Kura Araz	Lowlands	Game birds, mainly pheasants
Glinani island	400	Caspian, Kura-Araz	Wetland	Waterfowl, other birds
Byandovan	4,930	Caspian, Kura-Araz	Semi-desert meadows, wetlands	waterfowl, and plants of the Shirvan lowland
Korchay	15,000	Kura-Araz	Semi-desert meadows, wetlands	Persian gazelle conservation
Lachin	21,370	Lesser Caucasus	Mountain forests and meadows	Bezoar goat (Capra aegarus), bear, wild boar and hare conservation
Gusar	15,000	Greater Caucasus	Foothills and lowland forests	Game birds, wild boar and hare conservation
Shamkir	10,000	Kura-Araz	Wetlands	Game birds and waterfowl conservation
Absheron	815	Caspian sea	Wetlands	Waterfowl, Caspian seal conservation
Zuvand	15,000	Talysh	Mountain meadows and forests	Game birds, bear, leopard, and rare reptiles conservation
Ismaili	23,438	Greater Caucasus	Mountain and lowland forests	Rare and endangered species conservation
Gubadli	20,000	Lesser Caucasus	Mountain forests and meadows	Mammal conservation
Ordubad	27,869	Lesser Caucasus	Treeless mountain plateau	Mammal conservation
Kizil-Agach	10,700	Lenkeran	Wetlands	Waterfowl
Kiziljan	5,135	Lesser Caucasus	Mountain forests	Ecosystem conservation
Dashalti	450	Lesser Caucasus	Forests	Ecosystem conservation
Arazboyu	2,200	Kura-Araz	Riparian (tugay) forests	Forest conservation
Gabala	39,700	Greater Caucasus	Forests	Forest and rare species conservation
Gakh	39,700	Greater Caucasus	Fauna	Ecosystem conservation
Total area	**276,724**			

Sources: Chemonics International/USAID's Bureau of Europe and NIS, Biodiversity Assessment, 2000. Ministry of Ecology and Natural Resources, 2003.

Table 9.7: National parks

Name	Area in 2003 (ha)
Ordubad	12,131
Shirvan	54,373
Ag Gel	17,924
Total area	84,428

Sources : Chemonics International/USAID's Bureau of Europe and NIS, Biodiversity Assessment, 2000. Ministry of Ecology and Natural Resources, 2003.

First three National Parks – Shirvan, Ag-Gel and Ordubad – were created in 2003. Girkan, Alti-Agach and Gey-Gel strict nature reserves are being expanded. The necessary documentation for creating Absheron National Park is going through appropriate governmental procedures.

According to the National Report on the Ecological Network (2001) of the Committee of Experts for the Development of the Pan-European Ecological Network, submitted to the Council of Europe, the establishment of the national ecological network – AZECONET (specially protected natural areas) – is expected to be one of the most efficient measures for conserving biodiversity, because it would comprise the most important terrestrial ecosystems/habitats that need preservation.

The existing protected area system will be expanded by 500,000 ha to encompass types of habitats that are not sufficiently represented. This programme will run until 2010.

Another project on protected areas, which is important for sustainable rural development, is the Shah-Dag Rural Environment Project (see Box 9.1). It is supported by the World Bank, the Global Environment Fund (GEF) and other donors to the tune of US$ 13 million; the Government contributes US$ 1 million. The national focal point is the Minister of Ecology and Natural Resources. The project implementation unit was established in 1998. The Project's concept was recently revised, and the project appraisal document (approved in 2002) will be revised accordingly.

9.7   Information and awareness-raising

Stocktaking of birds is conducted twice a year; mammals – once a year. The Caspian marine survey on bioresources stock assessment is undertaken several times a year. The biodiversity strategy (UNDP) and Shah-Dag National Park (World Bank) projects envisage building databases and software. This would serve as a basis for biodiversity monitoring and significantly contribute to the development of a national information system and biodiversity management.

Box 9.1:   Shah-Dag National Park

Shah-Dag is the highest peak of the Northern Caucasus within Azerbaijan. The park will encompass Gusar, Guba, Divichi, Khachmaz, Ismaili and Shemkha districts (part of the territory between the Caspian Sea, the Russian border, and the Greater Caucasus Mountains). The core park area will be under strict biodiversity conservation, while the surrounding buffer zones will serve for sustainable economic activities that should support the sustainable rural development of local communities. The project addresses the key issues for improving biodiversity conservation:

• Integrated forest, rangeland and protected areas management. The park management plan will be developed in a participatory process, targeting particularly capacity-building of the public sector and improvement of environmental knowledge of local communities. It will be based on intersectoral cooperation and focused particularly on forest and tourist components and management.

• Rural energy supply and conservation. Measures for reducing unsustainable firewood use will be introduced by developing a rural energy strategy and public awareness campaign.

• Environmental management capacity-building and environmental education. This component will assist the Ministry of Ecology and Natural Resources to improve protected area management, forestry management and environmental management, including education programmes for civil servants in environmental situations

To ensure that the project is sustainable, appropriate incentives are planned to involve the local communities and demonstrate local benefits (e.g. local income-generating activities, such as ecotourism). The institutional and public capacity-building component is assessed as critical, because the Ministry, regional and park staff should maintain and replicate the project's outcomes once it has been completed. Pilot testing of the newly developed management tools and timely dissemination of the results to local communities are also intended to reduce the project's risks.

Several environmental NGOs are involved both in projects and in awareness-raising. For example, the Birdlife Organization works on high mountain ecosystem protection; the Ecological Education Organization is engaged in the development of the Caucasus mountain tourism route; and the Azerbaijan Society for the Protection of Animals cooperates with the World Wide Fund for Nature (WWF) to improve veterinary services, implement the Aarhus Convention, develop public education programmes on the environment and monitor the seals in the Caspian Sea.

All of the NGOs interviewed spoke about the difficulties involved in registering as well as in exercising all their legal rights, owing primarily to complex and time-consuming administrative procedures. (See chapter 3, on environmental information and public participation.)

9.8   Policy objectives and management

Policy framework

The President signed a decree on biodiversity protection in 2001. It stressed the importance of developing a biodiversity strategy.

To define conservation priorities, the Ministry started work on a national biodiversity strategy and action plan in February 2003. The project has received US$ 351,000 from the Global Environment Facility and technical assistance from UNDP. Experts from a number of different ministries (economic development, agriculture, transport, tourism, culture and construction, as well as environment), non-governmental organizations and scientific institutions also take part in its development.

As the first step to identifying biodiversity information needs, a questionnaire was prepared and distributed to the stakeholders. A flora database is already being set up, and a web site will follow. Other international organizations taking part in the development of the national biodiversity strategy and action plan, e.g. the World Bank, will support the establishment of a gene bank with a US$ 50,000 grant.

The National Programme for the Restoration and Expansion of Forests was approved in 2003 (see above). The Programme was developed to mitigate forest degradation by reforesting 30,400 ha

of forest land, afforesting 14,300 ha of non-forest land (to improve the sandy soils along water bodies, the coast and the transport network), and silvicultural works on 25,000 ha to support natural forest regeneration. It is financed from the State budget only.

Legal framework

The Law of Environmental Protection (1999) stipulates that the preservation of biological diversity is one of the basic environmental principles. According to this Law, the State has supreme powers and duties in defining biodiversity conservation policy, including:

- Issuing decisions on natural resource use, setting nature use limits and quotas and signing contracts as provided by law;
- Approving natural resource use regulations and permit issuance procedures for natural resource users;
- Approving and implementing conceptual plans and comprehensive programmes for the use, conservation and renewal of natural resources;
- Determining national inventory and national registration procedures for natural resources;
- Determining official monitoring procedures for the environment and natural resources; and
- approving the list of special environmental sites of scientific, environmental and biological importance, and establishing national reserves, national and nature parks, natural areas and other protected sites.

At the same time, the State is also responsible for:

- Designing natural resource use programmes, keeping inventories and monitoring;
- Enforcing laws and regulations on environmental, wildlife and plant protection and conservation, rehabilitation and use;
- Maintaining a national list of rare and endangered plant and animal species ("The Red Book");

The Law on Specially Protected Natural Areas and Objects (2000) defines nature protection categories, types of protected areas and related management objectives (research, monitoring, training, tourism). It lists the natural resource use and prohibited activities by protection category, and specifies the means of State control over them.

The Law on Wildlife (1999) defines general principles of wildlife management and lists the management tools, such as monitoring, habitat preservation, conservation of wintering and nesting grounds, and use forms and rights. Public use (except amateur and sports fishing) includes aesthetic, health improvement and recreational activities. Private use includes all types of use that involve hunting, provided a permit is obtained and the fee paid. According to article 48, fees are supposed to be used for reducing State expenditures for the study, preservation, reproduction and research of wildlife.

Should regulations on the use and preservation of wildlife in intergovernmental treaties that Azerbaijan has signed up to differ from those set out in this Law, the regulations of the international treaties shall be applied (art. 55). The appropriate executive bodies, municipalities and public organizations are responsible for the implementation of this Law.

The Law on the Protection of Plants (1996) deals with the protection of plants from pests, diseases and weeds. It establishes the State Service for Plant Protection with responsibilities for the sanitary control and use of chemical, biological and other forms of plant protection. The Law assigns certain responsibilities to enterprises, land users and citizens, and prohibits the marketing of agricultural products that violate legal requirements as to the content of toxic chemicals, nitrates and poisonous substances. Administrative and criminal penalties are also described in general terms.

The Forestry Code (1997) defines the State ownership of forests (all forests except the trees and shrubs on agricultural or private land), which will not be privatized. The right to use the forest fund depends on a lease agreement, forest cutting licence (for the cutting and transport of firewood), order (for collecting dead wood), and forest licence (for other forest users). The Code contains general forestry principles and lacks implementation mechanisms.

A law on hunting is currently under preparation.

International agreements

Azerbaijan ratified the Convention on Biological Diversity in 2000. The first National Report on Flora and Fauna was completed in 1996. The second report on agricultural animals and the third report on forest resources were drafted in 2003.

Azerbaijan ratified the Convention on International Trade in Endangered Species of Wild Fauna and Flora (CITES) in 1998. Permits from the Ministry of Ecology and Natural Resources are needed for the export of animals.

Azerbaijan ratified the Convention on Wetlands of International Importance Especially as Waterfowl Habitat (Ramsar) in 2001. Two wetland areas, one in the Gizil-Agaj State Reserve and another in the Ag-Gol State Reserve, were included on the Ramsar List.

Azerbaijan also ratified the Convention on the Conservation of European Wildlife and Natural Habitats (Bern, 1979) in 1999 and the Convention for the Protection of the World Cultural and Natural Heritage (Paris, 1972) in 1998.

Azerbaijan also participates in the Pan-European Biological and Landscape Diversity Strategy, which seeks to conserve ecosystems, habitats, species, their genetic diversity and landscapes of European importance through the development of the Pan-European Ecological Network.

Institutional framework

The Ministry of Ecology and Natural Resources was established in 2001, taking over the responsibility for biodiversity and forest management from the previous State Committee on Ecology and Nature Utilization Control and AzerForest Production Association.

Currently, three departments within the Ministry have nature conservation responsibilities:

(a) The Department of Biological Resources and Protected Areas deals with the implementation of international agreements, habitat conservation (e.g. Ramsar sites), species protection and the protected area system. Within the Department there are 14 nature reserves and 3 national parks. The Biodiversity Inspectorate is responsible for flora, fauna and hunting.

(b) The Fish Reproduction Department is responsible for setting fish quotas (national quota is defined by the Commission on Aquatic Bioresources of the Caspian Sea established in 1992). The Department issues fishing permits (together with quotas) to the companies certified for commercial fishing according to EU standards for equipment, technical and sanitary conditions (e.g. storage and cooling system). The fees are paid into a special account of the Ministry of Ecology and Natural Resources: US$ 3 per ton of sprat and US$ 30 per ton of sturgeon. Part of this money is allocated to fish protection and reproduction, and the rest goes to the Ministry of Finances. To export sturgeon, companies need a CITES permit from the CITES Management Authority affiliated with the Department of Fish Reproduction of the Ministry of Ecology and Natural Resources. The Fishing Inspectorate controls amateur and economic fishing in the Caspian Sea and inland waters. It coordinates five *rayon* offices, of which one is for inland waters and the other four for the Caspian Sea. They reported two types of illegal fishing: poaching sturgeon for illegal export and individual poaching for food. The latter is decreasing because private fishing companies are giving jobs to unemployed fishermen from the former State enterprise for fishing.

(c) The Forestry Development Department is the central authority for forest management. It coordinates 34 regional (*rayon*) enterprises for forest protection and regeneration (in addition to four regional nurseries and three enterprises for afforestation). These enterprises carry out sanitary cuts, protection works against pests and diseases, inspection, reforestation, seed collecting and nursery production. There are about 3000 permanent forest workers, of whom 250 are engineers. Additional workers are engaged for some seasonal forest work (e.g. reforestation).

In 2001, the President of Azerbaijan established the State Committee for Biodiversity of Genetic Resources. The Committee guides and supervises the development of the national biodiversity strategy and action plan, and approves the related documents. It consists of representatives of the Ministry of Ecology and Natural Resources, the Ministry of Agriculture, the Ministry of Economic Development, the Ministry of Education, the State Committee for Land and Cartography, and the National Academy of Sciences. The Deputy Prime Minister chairs the Committee. The national biodiversity strategy and action plan should be completed in 2004.

9.9   Conclusions and Recommendations

The Ministry of Ecology and Natural Resources has intensified international co-operation and initiated programs in order to improve biodiversity. Biodiversity management and forest management, however, lack inventories, which are the bases for monitoring the changes of habitats and status of species populations, and therefore for making appropriate decisions.

According to Article 5 of the Law on Environmental Protection, the powers of local governments are exercised as provided by law, but in further provisions these are not mentioned. The rights and obligations of citizens and civic organizations are generally stated (access to environmental information, compensation for damage, healthy environment, natural resources use and conservation, environmental protection, participation in environmental programmes, and demonstrations (arts. 6 and 7)), but the implementation mechanisms are not defined. The same applies to the Law on Specially Protected Areas and Objects, and the Law on Wildlife. In the latter, some provisions are very unclear, namely the expression 'the use of wildlife for the purpose of obtaining the products that comprise the basis of their vital activities' should be replaced with exact names of these products. Furthermore, bylaws are needed that specify the main actors, their responsibilities, funding schemes and time frames.

The national biodiversity strategy and action plan will set legal, institutional and implementation priorities. The project of the Shah-Dag National Park, as well as the three new national parks created by Presidential Decree in summer 2003, will define and test biodiversity management mechanisms. Implementation will depend on legal tools, and the lack of some implementation mechanisms could slow them down or undermine their outcome. It is particularly important to involve the local communities that are key stakeholders in sustainable biodiversity management.

Recommendation 9.1:
To improve the implementation of the Law on Environment Protection, the Law on Specially Protected Areas and Objects, and the Law on Wildlife, the Ministry of Ecology and Natural Resources should, as soon as possible, improve implementation mechanisms for biodiversity management (specifying the roles of all responsible

institutions at national, regional and local levels-including protected area managers- and related activities, sources of financing, and a time frame) and incorporate them into the current legislation.

The Programme on Restoration and Expansion of Forests was approved by President's Decree in February 2003, and the Ministry of Ecology and Natural Resources is in the process of establishing a forest inventory system. In addition, at the initiation of the Ministry and by decision of the Cabinet of Ministers, all industrial forest cutting has been suspended.

Recommendation 9.2:
The Ministry of Ecology and Natural Resources, in cooperation with other ministries, scientific institutions and non-governmental organization, should fully implement the Programme on Restoration and Expansion of Forests, by, inter alia:
- *Capacity building for forest administrators and other forest experts (additional university education and on-the-job training);*
- *Strengthening the University Forest Faculty, especially in inventory methods and new techniques;*

- *Improving the efficiency of forest inspection, and*
- *Intersectoral cooperation (primarily agriculture and tourism).*

The Caucasus is an important global centre for agro-biodiversity, which is also reflected in the landraces of crop plants and domestic animals that have been selected during the long period of agriculture in the region. The broad genetic variation found in many crops and their wild relatives, and also among domestic animals, need to be inventoried and conserved in Azerbaijan. There are a few attempts to support the conservation of traditional fruit varieties, but if more is not done, there is a risk that biodiversity will be irreversibly lost.

Recommendation 9.3:
It is recommended that the Ministry of Agriculture together with the Ministry of Ecology and Natural Resources should initiate discussions with donors and international organizations to establish projects that would contribute to the future conservation of landraces of crop plants and domestic animals. The promotion of landrace conservation should be included in the national biodiversity strategy and action plan.

Chapter 10

LAND USE, AGRICULTURE AND DESERTIFICATION

10.1 Introduction

Geographical areas, climate and land

About 60% of Azerbaijan is mountainous, and the rest lowland, mainly steppe and semi-desert. The mountain ranges of the Greater and Lesser Caucasus, and the Talish Mountains surround the Kura-Araz lowland and the Caspian coast. Forest covers 12% and agricultural land – mainly pasture and arable land – makes up 55% of the territory. The remainder, non-agricultural land, includes undeveloped land difficult to use but also land exploited for settlements, roads, and industry.

Azerbaijan has a broad range of climatic zones: from subtropical to temperate, and even alpine climate. The precipitation is mostly low – 60% of the land is arid or semi-arid – but varies from 200 mm on the southern coast of the Absheron Peninsula to 1200-1400 mm close to Lenkeran, further south. In the latter region the climate is humid and subtropical, and suitable for citrus and tea production. In other important agricultural districts, such as in the Kura-Araz lowland, precipitation is low, in particular during the hot summer. The climate is significantly colder in the high mountain ridges and mountain peaks of the Greater and Lesser Caucasus.

The increase in the Caspian Sea's level since the 1970s has resulted in land degradation and land losses on the coast. More than 800 km2 of land is now under the Sea, mainly in the vicinity of Neftchala, Lenkeran and Astara in the south.

Agriculture

Agriculture in Azerbaijan has a long history, and there is evidence that cereals were cultivated 6,000-8,000 years ago. Agriculture has been and is very important for the country's economy and its people. About 40% of the active labour force is employed in agriculture, and half the population lives in the countryside. Agriculture contributed 16.8% to the GDP in 2001, and is the second most important sector after the oil sector.

Until the dissolution of the Soviet Union in 1991, Azerbaijan was a supplier of products such as vegetables, tea, grapes and wine, citrus and other fruits, tobacco and cotton to other Soviet republics. These crops were cultivated on large State or collective farms using intensive farming techniques. Imports from other Soviet republics covered its need for cereals, meat and several other agricultural products.

Following independence in 1991, Azerbaijan's market links with other former Soviet republics were cut. The country was left with an agricultural specialization and products for which the demand was limited. Production of major products fell drastically, and the food industry collapsed. Deliveries of basic foodstuffs such as grain from other parts of the former Soviet Union ceased, which caused problems for food availability. The dramatic changes negatively affected the economy of the country and the living conditions of its population.

Agricultural production has recovered since 1998. In particular, the production of meat, eggs, cereals, potatoes and vegetables has increased. Markets for agricultural products and inputs are fully liberalized, and the effects are generally positive.

Azerbaijan has good potential as a supplier of fruits, nuts, vegetables and processed products such as wine and juice. At present oranges, apples, nuts, wine and apple juice are exported, but in small quantities. Poor product quality, and inefficient transport and conservation chains hinder export. The Russian Federation is an important potential market, but it has been difficult to establish regular deliveries across the border. Recently lorries with perishable products have been halted unexpectedly at the border.

A key aspect of the reform programme is land reform and farm restructuring, guided by the Land Reform Law adopted in 1996. World Bank projects and other projects have supported its implementation. According to the Land Reform Law, arable land and land with perennial crops totalling 1.7 million ha were transferred free of charge to rural inhabitants working in agriculture. Farmers were exempt from all taxes but the land tax for five years after privatization. Private land can be sold freely, inherited, leased and used as mortgage. The Land Market Law of 1999 provides for the sale and purchase of land, and allows for the sale of mortgaged land in cases of failure to repay credit.

The privatization of land as well as of animal production has been completed, and is generally considered to be a success. There are at present about 870,000 family farms. Only 40 State farms, mainly research farms, are still farming State land.

The average size of private farms is only 2 ha per family and subsistence farming is prevalent. This impedes agricultural growth. Land consolidation is now on the agenda as it is seen to be important for the development of a more efficient and market-oriented agriculture. There are some signs of consolidation. About 2,500 larger farms have been formed on the basis of voluntary agreements among individual farmers. There are also a small number of individual farmers with larger production units. Farmers' associations for cooperation on road building and rehabilitation of irrigation are developing.

Pasture and forest land has not been privatized. The municipalities own land, including 600,000 ha of pasture in the vicinity of villages. The remainder of the pasture and forest land is owned and managed by the State.

Pasture is divided into summer and winter pasture. The acreage of winter pasture is 1.8 million ha and of summer pasture 0.6 million ha (table 10.1). Winter pasture is in the lowland and lower parts of the mountains, frequently on desert and semi-desert vegetation, and summer pasture higher up in the mountains. Farmers lease the pasture either from the municipalities or from the State. A consequence of the conflict with Armenia is that summer pastures in Nagorny-Karabakh and surrounding districts in the Lesser Caucasus are not used. This affects about 60-70% of total summer pasture.

The lack of capital and credit opportunities, and the consequent low level of mechanization, worn-out machinery and low levels of agricultural inputs are negative factors for the sector. The lack of processing capacities for agricultural products is another significant constraint for the development of production, even if there are some new investments. A general trend, for example in the milling sector, is that small units are replacing large-scale processing. Another problem is that the new private farmers have limited access to information and technical advice.

Crop production

The acreage available for arable farming is 1.6 million ha. Arable crops are generally irrigated (85%) but there is also some rain-fed cultivation of, for example, cereals, potato and sainfoin (forage plant). Irrigation is or has been possible on 1.45 million ha. However, only 1.1 million ha can now be irrigated. This is one reason why only 1.162 million ha were sown in 2001; another is farmers' lack of resources and capacity. Yield levels are generally low even on irrigated land: average yields of irrigated cereals are about 2 tons per ha.

Important crops are: vegetables, cereals including rice, grapes, fruit, potato, cotton, tea and tobacco. Recent acreages for different groups of crops are found in table 10.1. Since 1990 the structure of crop production has changed considerably: cereal, vegetable and potato production has increased, while cotton, tobacco, fruit and grape production has plummeted. These trends reflect an adaptation to the needs of the national market and limited opportunities for export.

Table 10.1: Statistics on agriculture and land

Total land (1000 ha,2001)	8,641	*
of which agricultural land	4,740	*
of which forest	1,037	*
Agricultural lands includes:		
arable land (1000 ha, 2001)	1,835	*
of which irrigated land	1,097	***
perennial crops (1000 ha, 2001)	176	*
pasture and meadows (1000 ha, 2001)	2,670	*
Total pasture	2,384	***
Total winter pasture	1,780	***
Total summer pastures	604	***
Eroded land of total land resources	3,610	***
of which water erosion	2,972	***
of which wind erosion	368	***
of which irrigation erosion	269	***
Irrigated land subjected to salinisation		
(minsel, Gasiev)	612	***
of which weak salinisation	427	***
of which moderate salinisation	130	***
of which severe and very severe		
salinisation	55	***
Contaminated arable land	25	***
by oil and gas production	11	***
by building industry	5	***
by construction and waste disposal	2	***
by erosion and waterlogging	6	***
Sown area (1000 ha, 2001)	1,162	**
Cereals (1000 ha, 2001)	761	**
Industrial crops (1000 ha, 2001)	99	**
Potatoes, market garden crops and		
vegetables (1000 ha, 2001)	151	**
Fodder crops (1000 ha, 2000)	151	**
Animal production		
cattle (1000 heads, 2002)	2,098	**
cows and buffaloes (1000 heads, 2002)	1,002	**
sheep (1000 heads 2002)	6,003	**
goats (1000 heads, 2002)	556	**
pigs (1000 heads, 2002)	17	**
poultry (1000 heads, 2002)	15,351	**
Rural population of total (%), 2000	49	*
Employment in agriculture, hunting,		
forestry and fishing (% of total		
employment), 2000	41	*

Sources:

* Food Security in Azerbaijan 2001 (statistical yearbook).

** State Statistical Committee of Azerbaijan Republic. Environment. Statistical Yearbook. 2002.

*** Kosayev, E and Guliev, Y, Country Pasture/Forage Resource Profiles, Azerbaijan, FAO Grassland and Pasture Crops, 2001.

Animal production

Animal production is privatized and the animal stocks on State farms for breeding and other purposes are small. Traditional, extensive grazing dominates production. A few privatized large-scale production units for poultry and pigs still remain.

Recent animal stocks are accounted for in table 10.1. The share of animal production in agriculture declined during the first years after independence. Animal production is now quite profitable, and sometimes the only way for the rural population to make a living. From 1996 to 2002 the increase in animal stocks and production was substantial. Cattle stocks increased by 25%, sheep stocks by 35% and goat stocks by as much as 260%. The figures may be even higher than those reflected in official statistics. The decrease in pig stocks and pork production is an exception. The import of animal products has dropped sharply and Azerbaijan is now more or less self-sufficient.

Pastures and meadows meet as much as two thirds of the livestock's fodder requirement. More than 3 million head of sheep, goats and young cattle used to be taken to summer pastures in May-September. For various reasons summer pastures in the different mountain ranges are now used less intensively. The conflict with Armenia is one reason; another is the trend to keep the grazing animals in the lowlands and lower mountain regions as this is less costly. Frequently, hay is harvested from the alpine meadows and sold to farmers.

Manure is used as fertilizer, and is sometimes dried and used for fuel when there is no access to natural gas.

10.2   Environmental concerns

Irrigation and drainage

Irrigation is essential for agriculture in most parts of Azerbaijan and 41% of total water abstracted is used for irrigation. The main sources of water for irrigation are the rivers Kura, Araz and Samur (see chapter 7, on water management). Groundwater from drilled wells is also frequently used. The private farmers develop this source intensively as the major irrigation installations are falling into disrepair.

During the 1990s the infrastructure for irrigation and drainage deteriorated seriously, as only very limited funding was available for maintenance and rehabilitation. It is estimated that installations on 350,000 ha need major rehabilitation. Significantly less water is used for irrigation now than in the early 1990s. On the positive side, there are indications that land privatization has contributed to improved on-farm use of irrigation water.

There are three obvious environmental problems related to irrigation. Water ecosystems may deteriorate if too much water is abstracted. Irrigation and other uses of water have decreased the water flow of the Kura river with significant negative effects on the river ecosystem. Salinization of land and polluted run-off of drainage water are the other two main problems.

Losses in the distribution of water for irrigation are estimated at 50%. One reason is that few irrigation canals are lined with concrete. The water losses cause waterlogging and soil salinization. Another reason for the accelerating salinization is that only 600,000 ha of irrigated land, the most naturally saline areas, have drainage. In more than half the drained area the installations need rehabilitation. The increased water level of the Caspian Sea has also made land on the coast more saline. Salinization is particularly widespread on the Kura-Araz lowland.

Erosion

Soil erosion, in particular water erosion (table 10.1), is a widespread natural phenomenon in Azerbaijan due to the relief, soil conditions and climate. Wind erosion is sometimes a problem on the Kura lowland and the Caspian coast. Worsening erosion of unprotected coast is the result of the increased water level of the Caspian Sea. Development of mud volcanoes caused by gas leaks from underground hydrocarbon reserves is another natural process that contributes to erosion.

Erosion is accelerated by poor land management practices, such as ploughing of land on steep slopes; excessive cuttings of forests, shrubs and bushes; overgrazing; and irrigation washing away topsoil. Reduced dredging and riverbed management is intensifying the effects of seasonal flooding and hence the erosion of riverbanks.

Logging in the mountains, for firewood or other uses, is probably the most serious cause of increased water erosion. The lack of alternative energy resources, including dependable supplies of electricity, is one important reason for the extensive logging close to villages from the beginning of the 1990s. The conflict with Armenia seems to add to the pressure on forests in Nagorny-Karabakh and surrounding districts.

In the mountains, high pressure on some summer pastures has similarly intensified erosion processes.

Recent climate trends are also an important factor. The past years have been characterized by severe flooding of 10,000-15,000 ha and very high water flows in rivers, leading to intensive erosion and other damage to agricultural land. In 2003 there have repeatedly been unusually severe mudflows.

According to statistics, 1.3 million ha are mildly affected by erosion, 1.15 million ha slightly more so and 1.14 million ha are highly exposed. In all 3.6 million ha are affected by erosion, including 48.6% of agricultural land and 20% of forests.

Overgrazing and other consequences of animal production

Breeding of cattle, sheep and goats is currently profitable. This fact, coupled with the poverty of the rural population, is the main reason for the overgrazing that can be observed on much of the pasture. An analysis of animal stocks and the acreage of pasture available reveals that more animals (even according to official statistics) are grazing each hectare than is allowed.

It is unfortunate that sheep and in particular goat stocks have increased so significantly, as these animals tend to use the pasture more intensively. The most intensive use of pasture is found on the communal winter pasture close to the villages, belonging to the municipalities.

The large number of internally displaced people having moved with their cattle to arid parts in the lowlands contribute to the overgrazing of winter pasture.

Pasture is particularly degraded in the eastern Caspian region. The lower mountains of Gobustan and other low-altitude parts of the Greater Caucasus, and the lowlands suffer the most serious

degradation owing to a high pressure and suboptimal use.

Manure handling contributes to the pollution of water with nutrients and organic substances, probably with significant negative effects on drinking water.

Desertification

The arid and semi-arid climate in large parts of Azerbaijan is an important factor for desertification. Decreasing soil depth and changes in the plant cover are indications of desertification. The strong winds are another contributing factor, causing sand storms and severe desiccation of plants. A trend towards a drier and warmer climate has been recorded in several regions since the 1970s. The year 2000 was extremely dry . However, the main causes of the accelerated desertification are anthropogenic: cutting of forests, overgrazing and salinization of land.

Salinization, erosion, overgrazing and desertification are interlinked. Salinization of land is a severe problem in the lowlands, and water erosion in the mountains. Overgrazing can be a problem in the lowlands as well as in the mountains.

A study in the east of the country (40% of the total territory) revealed a considerable increase in desertification between 1982-84 and 1998. The study reported a doubling of the area with "intensive" and "very intensive" desertification from 23.5% to 45.1% of the investigated area. Despite this development, no special measures have been taken in recent years to combat desertification.

Desertification processes are most serious in the Absheron-Gobustan area, the Kura lowland and the lower mountains of Adzhinour, regions where agriculture is intensive. The rising level of the Caspian Sea, leading to salinization of coastal depressions, and intensive land exploitation on the coast are the main causes of desertification along the coast. Also on mountain steppes, previously often covered by forests, the land is sometimes used and overused for grazing and hay production with desertification as a consequence. About 30% of all land in Azerbaijan shows different degrees of desertification.

Many of the sensitive areas of the Kura-Araz lowland and the Absheron Peninsula are densely populated: 5 million of the country's 8 million people live here. Different human activities, such as transport, particularly off-road, add to land degradation and desertification. Industry and oil exploitation also contribute to desertification mainly on the Absheron Peninsula. Also on the Absheron Peninsula, the exploitation of sand and limestone as building material has a negative impact.

Fertilizers and pesticides

During the Soviet period, high quantities of pesticides were used on crops such as cotton. DDT was used regularly in the 1970s and 1980s, although it was officially prohibited. Excessive application of pesticides and chemical fertilizers has polluted soil and groundwater, and Azerbaijani scientists have linked this pollution to birth defects and illnesses.

Pesticide use has declined dramatically from 87 kg/ha in 1990 to 5 kg/ha in 1997. The use now seems to have stabilized on a still lower level. According to official statistics about 500 tons of pesticides are imported annually. Moreover, some pesticides seem to be imported illegally. The use of fertilizers is also very low.

The lack of training and advisory services for farmers hampers the proper selection and use of pesticides and fertilizers. Moreover, pesticide sprayers are outdated and do not lend themselves to an even distribution of the active substance. The same problem is true for the spreading of fertilizers. Farmers rarely protect themselves when applying pesticides.

Agricultural run-off, in particular nutrients, contributes to the pollution of rivers, groundwater and the Caspian Sea (see chapter 7 on selected Caspian sea issues). The additional run-off caused by the use of fertilizers and pesticides is, however, most likely to be low due to their limited use. As there is no systematic monitoring, this cannot be confirmed. There are instances where, for instance, DDT is still found in run-off, but it cannot be concluded that contamination with persistent pesticides such as DDT is still a problem.

Land contamination

In the vicinity of Sumgayit, soil is contaminated by mercury, fluorine and benzopyrenes. Other pollutants are also emitted, and heavy metals and persistent organic pollutants accumulate in soil and food in the area.

On the Absheron Peninsula, about 10,000 ha are contaminated by oil, which also has a negative effect on the natural biotopes and contributes to desertification.

Landmines and unexploded ammunition are a problem in certain regions. Landmines are found mainly on land affected by the conflict with Armenia, and ammunition in the vicinity of former Soviet military bases. (See Chapter 12, on human health and environment.)

10.3 Policy objectives and management

Policy framework

Two of the five main problems identified in the National Environmental Action Programme (NEAP) of 1998 are linked to agricultural production and land degradation: degradation of agricultural land, loss of forestry and biodiversity; and water quality. The NEAP conclusions are also reflected in the National Programme for the Restoration and Expansion of Forests (2003, see chapter 9, on biodiversity and forest management).

The recently adopted National Programme on Environmentally Sustainable Socio-economic Development (2003) includes the following measures with regard to land resources:
- Development of a national action plan to protect agricultural land and increase its fertility;
- Implementation of appropriate measures to prevent erosion and improve saline soil;
- Continuation of recultivation works on oil-contaminated land, giving priority to setting up facilities to clean up the soil;
- Inventory of existing drainage, collector and irrigation networks and their improvement;
- Regular monitoring of soil which has lost fertility owing to mining, and underground and surface construction, and implementation of appropriate measures for recultivation; and
- Implementation of measues to overcome the adverse environmental effects of sludge around industrial centres (Sumgayit, Ganja).

The Programme also identifies a number of measures to be implemented in the agricultural sector, some of which are linked to land management:

- Enhanced use of optimized planting systems and advanced irrigation technologies to prevent erosion and salinization of agricultural land;
- Development of national and regional schemes for the use of chemicals in agriculture; and
- Prevention of the use of technologies that may pollute and degrade soil.

Azerbaijan ratified the United Nations Convention to Combat Desertification in 1998, but no action programme under the Convention has yet been developed. It is expected that GEF will make some funding available for the implementation of an action programme. The National Programme on Environmentally Sustainable Socio-economic Development includes the following measures to combat desertification:

- Inventory of land affected by desertification;
- Development and implementation of a national action plan against desertification;
- Development of a system of forecasting and awareness to improve the effectiveness of anti-desertification measures; and
- Planting of forests.

Another important recent initiative, the State Programme on Poverty Reduction 2003-2005, includes support for the development of farmers' unions, improvement of local advisory services, rehabilitation of drainage and irrigation, and further development of water users' associations.

A national programme for the rational use of winter and summer pasture, meadows, and stopping desertification is being developed by the Ministry of Ecology and Natural Resources in collaboration with the Ministry of Agriculture. A draft of the programme made available during the Environmental Performance Review mission was very broad in terms of its different long-term and short-term activities. A national programme for soil conservation is being prepared by the Soil Conservation Commission under the Ministry of Ecology and Natural Resources.

Over the past decade only marginal resources have been devoted to monitoring erosion and desertification. This is to some extent in contrast with the Soviet period, when this and other land-use problems were better studied and analysed.

Legal framework

In recent years several important laws and regulations have been adopted. The Land Reform Law of 1996 was discussed above.

The Land Code was adopted in 1999. It gives the framework for the division of land into different categories and regulations for its use. Regulations for the use of pasture are an important part of the Land Code. The Cabinet of Ministers makes decisions to lease out State-owned pastures to physical or legal persons, in agreement with the *rayon's* executive power, the State Committee for Land and Cartography and the Ministry of Ecology and Natural Resources. Criteria regarding the maximum number of animals per hectare are set. Municipalities are responsible for leasing out part of their pasture, while a proportion is kept for communal use. The period of leasing is 10-99 years (State land) or less than 15 years (municipal land). A presidential decree from 28 November 2000 determines the controlling function of the State Committee for Land and Cartography with regard to land use and protection, including leased land.

Three other laws on land use are also from 1999: the Law on Leasing of Land, the Law on the Land Market and the Law on the State Land Cadastre, Land Monitoring and Land Survey. A number of regulations under these legal acts have also been adopted.

The 1996 Law on Plant Protection gives the legal framework for the use of pesticides. The 1997 Law on Pesticides and Agrochemicals regulates the testing and registration of pesticides and agrochemicals, and defines the organization of agrochemical services.

The Law on the Preservation of Soil Fertility was adopted in 1999. It gives the legal basis for the protection of State, municipal and private land, and the protection of soil fertility. The competent authorities are the Ministry of Agriculture, the Ministry of Ecology and Natural Resources, the State Committee for Land and Cartography and the Committee for Amelioration and Water Management.

Economic instruments and financing

A system of water users´ associations is being put in place with the support of the World Bank. Individual farmers sign a contract with their associations and pay them for irrigation water. The associations pay water fees corresponding to 8,000-18,000 manats/ ha to the Committee for Amelioration and Water Management. The farmers pay the same amount to the associations plus 25% to cover the costs of the latter. The payment for water covers only 8% of the running costs for the upkeep of the infrastructure.

The lack of funding is one reason why it has been difficult to implement agricultural policies. International projects are important for the development and implementation of agricultural and land management policies.

Projects funded by USAID give opportunities for farmers to obtain microcredits and support the development of structures selling agricultural inputs. Loans from the World Bank are also used to establish microcredit lines for farmers, to develop extension services and to rehabilitate parts of the irrigation infrastructure.

The EU Food Security Programme has been ongoing since 1996. The total assistance of €49 million is used mainly as direct budget support for the Ministry of Agriculture for policy reform with regard to food security and poverty reduction, market liberalization, and land reform.

In a project funded by the World Bank/GEF, the Shah-Dag Rural Environmental Programme (see chapter 9 on biodiversity and forest management), one component is improving pasture management and finding mechanisms to improve leasing arrangements for forest and pasture.

Institutional framework

The four key authorities responsible for different aspects of agricultural policy and land management are the Ministry of Agriculture, the Ministry of Ecology and Natural Resources, the State Committee for Land and Cartography, and the State Committee for Amelioration and Water Management.

The Ministry of Agriculture has the primary responsibility for agricultural policy development. It includes the State Veterinary Committee and the State Quarantine Committee. The Department for the Control of Land Tenure, Amelioration and Protection of Nature has a specific responsibility for environmental and natural resource issues within the Ministry. The State Chemical Commission under the Ministry of Agriculture is responsible for testing and registering pesticides. The Ministry has been reformed in recent years to adapt to the privatization of agricultural production. The reforms have resulted, among other things, in a drastically reduced staff.

The Ministry of Ecology and Natural Resources and its regional centres are in charge of the implementation of environmental legislation including on-farm handling of pesticides, fertilizers and manure. It is also responsible for the management of forests (chapter 9 on biodiversity and forest management). The Soil Conservation Commission is subordinated to this Ministry.

The State Committee for Land and Cartography is the result of the merger of the Land and Cartography Committees in 2001. It has been responsible for the implementation of land reform. Its functions include administration of land titles, cadastres and land mapping, soil erosion control and implementation of measures to control salinity as well as pastures management.

The Committee for Amelioration and Water Management is in charge of land improvement activities on irrigated land, and the operation and maintenance of the irrigation infrastructure (see chapter 7, on water management.)

Cooperation and coordination between these key authorities is not very well developed. There are further overlaps in their mandates with regard to:
• Land-use policy and regulations;
• Management of pastures;
• Implementation of soil conservation measures; and
• Water policy and management of irrigation.
As a result it is difficult to make decisions, implement measures and enforce existing legislation.

Local farmers' associations to carry out specific tasks are developing, but there is no national farmers' union. Some NGOs in the agricultural sector are involved in advisory and other activities.

Extensive research capacities are found in various institutes of the Academy of Sciences and the Agrarian Academy, but the whole research and education system is under severe pressure owing to a lack of funding. The Ministry of Agriculture recently established ten regional agro-scientific centres for research and extension.

10.4 Conclusions and recommendations

The difficult social and economic problems are the main reason why farmers pay little attention to environmental issues. Even a severe threat to their future production capacity such as erosion is not given enough attention. Where there are no alternative energy sources to firewood, even forests planted to protect against erosion are being cut.

Land degradation is one of the most serious environmental issues in Azerbaijan. Processes such as erosion and desertification seem to be accelerating, which is distressing as they are to a large degree irreversible.

The Ministry of Agriculture, the Ministry of Ecology and Natural Resources, the State Committee for Land and Cartography, and the State Committee for Amelioration and Water Management are the four national authorities responsible for different aspects of agricultural policy and land management. Overlapping functions between these authorities and unclear mandates make decision-making and implementation of decisions related to land management difficult. As a result the very limited financial and human resources are used inefficiently. It is essential that each authority should have its own specific role and responsibility.

One option is to give the Ministry of Ecology and Natural Resources overall responsibility for planning and control with regard to land use and land conservation, to focus the responsibility of the State Committee for Land and Cartography on land markets and transactions, mapping and cadastres, and to merge the Ministry of Agriculture with the State Committee for Amelioration and Water Management. The merged ministry's main tasks would be to develop sustainable agriculture, agricultural markets, food security and services to the sector. The use of agricultural land and the development of irrigation would be important responsibilities for this new ministry.

In a reformed system of land management, it is important to involve and give more responsibility

to the rayon, municipal and community levels. On the rayon level the different agencies involved should be made to cooperate under the governor. The municipal level, already important, could be even more influential.

Recommendation 10.1:
The Cabinet of Ministers should appoint an interministerial working group to review and rationalize the responsibilities for land management of the Ministry of Ecology and Natural Resources, the Ministry of Agriculture, the State Committee for Land and Cartography and the State Committee for Amelioration and Water Management as well as the rayon *and municipal authorities. Among the issues to be resolved are the following:*
- *Assignment of responsibility for an information system on land and land degradation; and*
- *Development of a strategy for land conservation and sustainable land use.*
(see Recommendations 10.2 and 10.3).

As has been stated in other chapters of this review, the lack of monitoring data and of recent inventories seriously impedes the development of cost-efficient policies and action programmes. This is a matter of concern for the authorities involved in land management. For an important problem such as land degradation, more resources must be devoted to gathering and analysing information.

The establishment of an information system on land and land degradation should be made in an integrated way, for all land types and for all processes such as erosion and desertification. A step-wise approach is recommended, with information necessary for the evaluation of the most serious problems a priority.

Recommendation 10. 2:
Based on the decisions of the Interministerial Working Group (see Recommendation 10.1), the responsible body should develop an integrated and unified database on land and land degradation as direct support to the development of a strategy for land conservation and sustainable land use, and land management projects and programmes. The database should be accessible to all authorities and other stakeholders in land management and land conservation.

With the exception of the reforestation programme, Azerbaijan has so far made only very marginal efforts to counteract land degradation. However, there are a great number of activities and programmes included in recently presented initiatives. While land degradation is a complex process with clear linkages between, for example, erosion, desertification, and overgrazing, it has not been possible for Azerbaijan to give equal weight to all of these issues.

One important aspect of action programmes for land management is that many activities are expensive, e.g. the rehabilitation of drainage and terracing of steep slopes. The limited resources available are an additional reason why an integrated and focused approach is necessary.

It is essential that future programmes should build on the involvement of farmers, communities and municipalities. Cost-sharing and contributions in kind from these stakeholders should be a longer-term objective.

An integrated approach does not exclude that actions are prioritized. Communal pasture, for example, is under severe pressure. This category of land and other overgrazed pasture needs immediate attention.

The lack of knowledge about the impact of various possible policy instruments hinders the improvement of land management. In the medium term it is essential to know which policy instruments are the most efficient.

Recommendation 10.3:
Based on the decisions of the Interministerial Working Group (see Recommendation 10.1), the responsible body should:
(a) Develop a prioritized and integrated strategy for land conservation and sustainable land use; and
(b) Derived from this integrated approach, develop targeted programmes for priority issues, for example, for combating desertification or improving pasture. Projects should be developed to test different policy tools.

So far organic farming is not on the agenda in Azerbaijan, and it will take a long time for organic farming to develop. However, developing organic farming is important when moving towards sustainable agricultural production. Market niches for organic products, including export products such as fruit, wine and tea, could increase the income of the producers significantly.

The main bottleneck at this stage is the need to inform and develop the interest of farmers and consumers in organic farming. At a later stage a labelling system for certified products will need to be developed.

Recommendation 10.4:
The Ministry of Agriculture should promote the development of organic farming and eco-labelling of food products. Support should primarily be directed towards capacity-building and the establishment and development of organizations for organic farming.

It is understandable that environmental issues are not the primary focus of attention for the rural population and the farmers. The rural population may be more or less compelled to overexploit forests and pastures for their subsistence. This is the reason why, also from an environmental perspective, it is important now to support the general development of the agricultural sector and the rural economy.

Azerbaijan has been quite successful in organizing extension services, which are a key instrument in the development of agricultural production. Farmers need support in their new role as independent farmers to find ways to earn a living and to develop their production. Some printed information material has been developed and distributed, but more needs to be done. Institutions in direct contact with farmers are the rayon offices of the Ministry of Agriculture, agricultural institutes, the 10 regional agro-scientific centres recently established by the Ministry of Agriculture, and NGOs. Private sector advisory services are also being set up in the framework of a World Bank project. The efforts made to develop the possibilities for farmers to get information and training are positive, but should be strengthened.

Recommendation 10.5:
The Ministry of Agriculture, in the longer term, should encourage the extension services to implement codes of good agricultural practices, including supporting the farmers to establish nitrogen management plans or apply integrated pest management. In this respect it is important to have a scientific basis and to make efforts to safeguard basic needs.

PART III: ECONOMIC AND SECTORAL INTEGRATION

Chapter 11

ENVIRONMENTAL CONCERNS IN THE OIL AND GAS SECTORS

11.1 Introduction

Azerbaijan is the cradle of the oil industry, both onshore and offshore, and its environment has paid part of the price for this pioneering role. In countries with a market economy, the externalities of oil and gas activities were realized some decades earlier than in the former socialist countries. As the countries with economies in transition came to understand their environmental problems soon after their independence, they were also facing an economic crisis, leaving scant attention for environmental issues. The same is true for Azerbaijan.

The history of drilling for oil and gas in Azerbaijan stretches over more than 130 years. In the second half of the 19th century, the area that is now Azerbaijan was the world's main oil producer alongside the United States. Shallow water extraction started in the 1920s and moved into deeper waters in the late 1940s. Despite the promising prospects, the former Soviet Union concentrated on exploration and development in areas that held more easily accessible reserves. This left Azerbaijan's resources largely unexploited until now.

There was sufficient exploitation, however, to cause environmental problems. The environmental performance of the technology used for early drilling onshore in Azerbaijan was poor, as was the case in the rest of the world. This resulted in oil spills and production leakages, which seriously polluted the land and are still visible today. The pollution was compounded by associated water, drilling mud and drilling cuts contaminated by drilling mud. These substances have been poured into open ponds and streams. Natural seeps have also contributed to the contamination. Almost all of the contaminated areas of Azerbaijan have

been left untouched, and production in these areas continues. The level of contamination could be severe, particularly since there has also been uncontrolled disposal offshore.

The oil and gas sector of Azerbaijan is also the key sector of its economy. It is the engine that pulls and pushes the economy and will most likely be the most important economic sector for some considerable time to come. Azerbaijan is obviously in a proven oil province and the resources are known to be substantial, although the area has not been fully explored. The main part of the resources is believed to be in the subsoil of the Caspian Sea.

Everywhere the exploitation of natural reserves is both an economic and a political issue, and Azerbaijan is no exception. It is the world's oldest producer, and the industry has been a more dominant part of the economic activities than anywhere else. The oil industry in Azerbaijan is therefore also the sector of the economy that is often the most attractive to scientists and other experts.

11.2 The oil and gas sector

Territory

The oil and gas reserves of Azerbaijan are both onshore and offshore. Although the Caspian Sea is a proven oil and gas area and production has been going on for years, the issue of territorial waters has not yet been fully settled among all littoral countries. Talks are ongoing and the majority of countries agree on the principles for drawing the borders, but there are outstanding claims from two countries against some of the proven reserves that today are considered to be in the subsoil of Azerbaijan.

Figure 11.1: Areas covered by Production Sharing Agreements, name of PSA and operator.

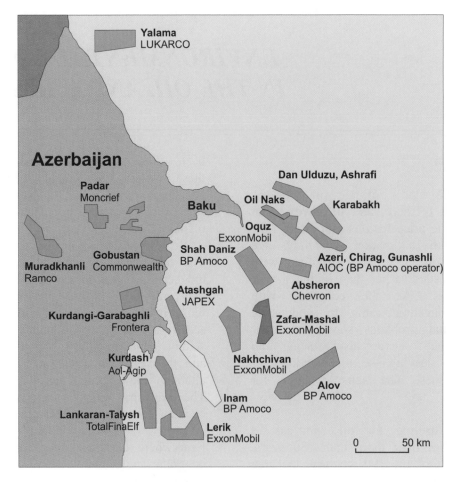

The subsoil of the Caspian Sea is considered to very rich in oil and gas reserves. Offshore production in relatively shallow waters has been going on for many decades. The exploration activities are organized as production sharing agreements (PSAs), all with the participation of Azerbaijan's State Oil Company (SOCAR) and a consortium of other oil companies, including BPAmoco, ExxonMobile, ChevronTexaco Agip, Total and Lukoil, among others. Since 1994 Azerbaijan has signed 21 PSAs and 2 contracts for pipeline construction with groups of companies (one of the pipeline contracts is combined with a PSA). Fourteen of the PSAs are offshore; five are onshore. Thirteen of the offshore PSAs are in the waters south of the Absheron Peninsula, and only one to the north.

The waters south of the Absheron Peninsula are in an area with proven oil and gas reserves and are therefore also very attractive for the oil companies. To the north of the Absheron Peninsula only one PSA has been signed, opening the area for exploration and exploitation. This area is likely to become much more attractive.

The orography of the Azerbaijani part of the Caspian seabed is a steep increase in water depth from the coast north of the Absheron Peninsula, and the same from the coast to the south, with water up to 1000 m deep. From the Absheron Peninsula stretches a sub-sea ridge with relatively shallow water all across the Caspian Sea. It is on this ridge that large fields were found and they will now be developed. It is the mid-part of the ridge, holding some proven resources, that is disputed by another country. The geology of the Caspian subsoil is very complex, and the rigs that are drilling need to be equipped for handling pressures higher than the norm.

Oil and gas resources

The estimated proven oil reserves of Azerbaijan vary from 4 to 14 billion barrels (approximately 800 million to 2,800 million tons). This range is large and systematic work on the estimates could narrow it. A recent World Bank study conservatively estimates 4.5 billion barrels, equal to 900 million tons. There is no official estimate. Whatever the figure, the reserves are expected to grow during the coming decades due to the exploration already planned. The oil and natural gas will create a significant inflow of capital to Azerbaijan, which could support the development of an economic structure that could last beyond the oil and gas reserves.

The proven reserves of natural gas were 800 billion m³ until the gas and condensate field Shah Deniz was discovered in 1998. The field is being developed and the estimates of the proven reserves are now around 1,500 billion m³.

Oil and natural gas production and development

Oil

The production of oil from 1992 until 2002 is shown in Figure 11.2 below.

Azerbaijan's oil production (total liquids) averaged 318,000 barrels a day (bbl/day) in 2002, of which approximately 310,000 barrels were crude oil. This represents a 4% increase over 2001, and builds upon five consecutive years of growth (see Figure 11.2). At the same time, domestic petroleum consumption has decreased since independence, resulting in a growing margin for net petroleum exports (see

Figure 11.2). Azerbaijan exported approximately 178,000 bbl/day in 2002, most of which was routed to the Russian Federation, Turkey and Italy.

Most of the SOCAR production (65% in 2001) is derived from the offshore field "shallow-water Gunashli," located 100 km off the Absheron Peninsula. Shallow-water Gunashli first came on-stream in 1979, but was developed only to a maximum water depth of 120 m owing to technological constraints. Although foreign firms have shown interest in developing this site, no agreements have yet been reached. While production levels have been stable over the past five years, press reports suggest that the structure may be losing reservoir pressure.

In August 2002, SOCAR began to rehabilitate shallow-water Gunashli independently by adding a new production platform designed to enable 12 new wells to operate. SOCAR also operates 40 other older fields (both onshore and offshore), many of which are considered to be in similar disrepair and have been artificially stimulated for years using water injection. Press reports indicate that inefficiencies owing to ageing equipment and largely depleted reservoirs have caused the cost of production of the SOCAR onshore crude oil to reach some US$18 per barrel.

Figure 11.2: Oil production and consumption 1992 – 2002

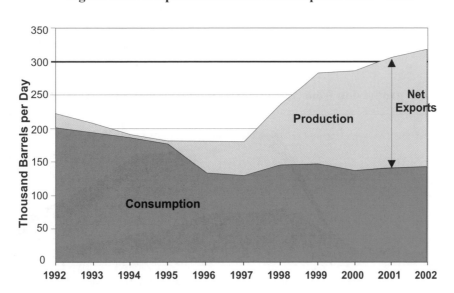

Source: EIA

The Azeri-Chirag-Gunashli (ACG), Shah Deniz, Baku-Tbilisi-Ceyhan (BTC) pipeline and South Caucasus pipeline (SCP) projects are set to increase Azerbaijan's oil and gas production and export using modern technology and applying international standards to production, pipeline operations, maintenance, control, and environmental, health and safety management. The projects should generate very substantial revenues for Azerbaijan and a steady income flow from transit fees for Georgia and Turkey. They will also contribute to gas supplies in all three countries. (See section, below, on gas pipelines.)

The ACG and Shah Deniz projects will be developed gradually. Output from the ACG field is forecast to rise to over 500,000 bbl/day by 2007 and to about 1 million bbl/day (50 million tons a year) by 2009. Between 2005 and 2020, Shah Deniz is expected to add 110 million tons of oil equivalent (toe) to Azerbaijan's gas production, which stood at 4.8 million toe in 2000. (Figure 11.3)

The production forecast shown above is for the BP-operated PSAs, and from approved development activities only. This production is on top of the approximately 175,000 bbl/day produced from pre-1997 facilities, but it is expected to decrease over the coming decades.

The estimate is based on the proven reserves and reserves which Azerbaijan has decided to develop and produce. It is expected that these reserves are likely to increase as a result of new exploration.

Natural gas

Despite boasting one of the world's largest natural gas field discoveries of the past 20 years in Shah Deniz (see figure 11.4), Azerbaijan is currently a net natural gas importer. The country produced 200 billion cubic feet, or approximately 6 billion m^3, of natural gas in 2001, while consuming roughly 240 billion cubic feet, or approximately 7 billion m^3. Most of its natural gas imports currently come from the Russian Federation.

Azerbaijan has proven natural gas reserves of roughly 85 billion m^3, and potentially even greater reserves. However, because there is insufficient infrastructure to deliver the country's natural gas from offshore fields (the source of most of its production), natural gas has been flared off instead of being piped to markets.

Virtually all of Azerbaijan's natural gas is produced by SOCAR from offshore fields. The country's leading natural gas producer, the Bakhar oil and gas field, is located off the southern tip of the Absheron Peninsula and currently accounts for slightly over half the country's natural gas output. Output at Bakhar has been declining in recent years and, according to press reports, SOCAR has begun to develop a new deposit, known as Bakhar-2, near the existing Bakhar field. SOCAR reportedly has plans to route some of the Bakhar-2 natural gas production for export in the near future. In addition to SOCAR, the Azerbaijan International Operating Company (AIOC) and other international consortia also produce small amounts of associated natural gas.

Figure 11.3: Oil production from Azeri-Chirag-Gunashli and Shara Deniz 1998 to 2024

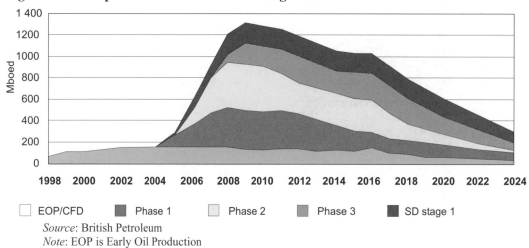

Source: British Petroleum

Note: EOP is Early Oil Production

Figure 11.4: Natural Gas Production and consumption 1992 to 2001

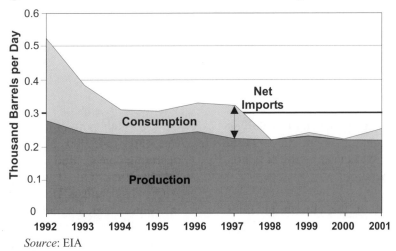

Source: EIA

Azerbaijan's natural gas production is expected to increase with the development of the offshore Shah Deniz field, which contains potential recoverable reserves of roughly 400 billion m^3 of natural gas according to the project's operator, British Petroleum. However, other industry and trade sources, using widely different definitions of "reserves", estimate the field's size to be as high as 1,000 billion m^3. Shah Deniz is located in the Caspian Sea, approximately 100 km southeast of Baku (see Figure 11.1), and is being developed by the Shah Deniz consortium (members: BP, Statoil, SOCAR, LukAgip, NICO, TotalFinaElf and TPAO). The Shah Deniz production sharing agreement was signed in 1996.

The first phase of the Shah Deniz field's development was officially sanctioned on 27 February 2003 and has been estimated to cost over $3 billion. It entails the installation of a new fixed offshore platform, two sub-sea pipelines to bring the hydrocarbons ashore, and a new onshore gas-processing terminal to be built close to the existing oil terminal at Sangachal, near Baku. Shah Deniz consortium expects to begin producing natural gas for export in 2006. According to the partners, once the new infrastructure is in place, Shah Deniz will be capable of producing approximately 296 billion cubic feet of natural gas (8.4 billion m^3) a year (roughly on a par with Bahrain's natural gas production in 2001) as well as 40,000 bbl/day of condensate. Natural gas from Shah Deniz, as well as associated gas from ACG and the Bakhar-2 project, is expected to make Azerbaijan self-sufficient in natural gas and to generate significant export revenue.

Refining

Azerbaijan's crude oil is refined domestically at two refineries: the Azerineftyag (Baku) refinery, with a capacity of 206,400 bbl/day, and the Azerneftyanajag (New Baku) refinery, which has a capacity of 137,600 bbl/day. With domestic production averaging 318,000 bbl/day in 2002 (and half of that exported as crude oil), Azerbaijan's refineries have been running well below capacity, with overall refinery utilization rates as low as 40%. Heating oil accounts for roughly half their output, followed by diesel fuel, petrol, motor oil, kerosene and other products.

Both refineries are in need of modernization, which the Azerbaijan Government estimates will cost between US$ 600 million and 700 million. The United States Trade and Development Agency will finance a $600,000 feasibility study for upgrading the two refineries and the specialized oil port of Dubendi.

The current refining capacity of Azerbaijan is approximately 344,200 bbl/day, which is approximately double the country's oil production in 1997. SOCAR runs both refineries.

Much of the Azerneftyag refinery was constructed in the early 1930s, though the site dates back to the 19th century. It contains several crude and vacuum distillation units, though it lacks the reforming units necessary to produce marketable petrol. Currently, it is used mainly to produce lubricating oil and bitumen, with diesel oil, fuel oils and straight-run kerosene as by-products. However, its facilities use old technology unable to meet modern requirements for lubricants, which are characterized by a high viscosity index.

Baku's second oil refinery, Azerneftyanajag (former Novo-Bakinsky), was constructed in 1953 and substantially rebuilt in 1965. It contains a fluidized catalytic cracker, a coker and several light-end polymerization units. A reformer was installed in 1980, and a combination catalytic cracking unit was brought on-stream in 1993 to improve the quality of high-octane petrol. It is not, however, currently in operation.

The low technical level of the current facilities and equipment plagues both refineries. The losses at the refineries are four to five times higher than at a modern refinery and stand at 2-2.5% of the processed oil. Refineries also consume two and a half to three times more energy and two to three times more chemicals than a modern refinery of the same capacity. Even a low purchase price for crude oil cannot ensure an acceptable level of income for the refineries, especially for the older Azerneftyag. The low technical level also significantly impacts the quality of the resulting products, few of which correspond to international standards.

Transport

The production of oil and gas requires transport to reach the markets. If quantities are large and planned over a considerable number of years, pipelines are in all senses the most efficient means of transport. Azerbaijan as a proven oil province is connected to oil pipelines. One connects it to the Russian Federation, and another, through Georgia, connects it to the harbour facilities of Supsa on the Black Sea coast.

Oil pipeline

The northern pipeline (No. 1) (figure 11.5) connects Baku to Novorossiysk in the Russian Federation and has a capacity of 103,000 bbl/day of crude oil. It has occasionally been out of operation as it passes through the conflict area in Chechnya. The Russian pipeline operator, Transneft, built a 120,000 bbl/day Chechnya bypass and restarted operation in December 2000. However, the northern pipeline has an oil quality problem, as the Russian Federation transports low-quality heavy crude through the same pipeline, and this gets mixed with Azerbaijan's high-quality light crude, reducing its value. SOCAR estimated the

loss of revenue in 2001 due to the pipeline problem at US$40 million to 50 million. Currently, SOCAR transports 50,000 bbl/day via the northern pipeline. AIOC has chosen to transport all its oil via the western pipeline instead, to ensure that the value of the oil is not reduced.

The western pipeline (No. 2) connects Baku to the Georgian port of Supsa on the Black Sea; it can carry 145,000 bbl/day, and it is currently operating near full capacity. The Georgian International Oil Company, a subsidiary of AIOC, has built a US$ 550 million oil terminal in Supsa and upgraded the 800-km pipeline. In 2001, AIOC exported 130,000 bbl/day of Azerbaijani oil via this pipeline.

However, export capacity using the Supsa terminal is limited not only by the capacity of the pipeline, but also by the capacity of tankers in the Bosphorus Strait. The constraints of passing through the Bosphorus Strait seem to be among the decisive factors for the Baku-Tbilisi-Ceyhan (BTC) pipeline.

Construction of the BTC pipeline (No. 3) is based on a contract between Azerbaijan and a large number of companies. This pipeline will link Baku with the Turkish port of Ceyhan on the Mediterranean Sea via Georgia. The pipeline will have a capacity of about 1 million bbl/day. In addition, it could carry oil from Kazakhstan westward, if the proposed trans-Caspian pipeline (No. 5) is built. Current plans are for the BTC pipeline to begin transporting oil in early 2005. Investors in the BTC pipeline include most of the companies that have also invested in AIOC, and it is expected that they will finance 30% as equity (or individual loans), and the remaining 70% will be financed by the Export-Import Bank of the United States, Japan's Export-Import Bank, the International Finance Corporation and the European Bank for Reconstruction and Development (EBRD).

Before the BTC pipeline gathered general support, a 300-km pipeline from Baku to Tabriz in the north-west Islamic Republic of Iran had been proposed (No. 4). It would have connected to the existing Iranian pipeline network and fed Iranian refineries. Azerbaijan, however, did not consider this to be the best option for a number of reasons, including the greater carrying capacity of the BTC pipeline, differences in oil quality and a preference to avoid dependence on another country's pipeline system.

Figure 11.5: Existing and proposed pipelines for export of oil

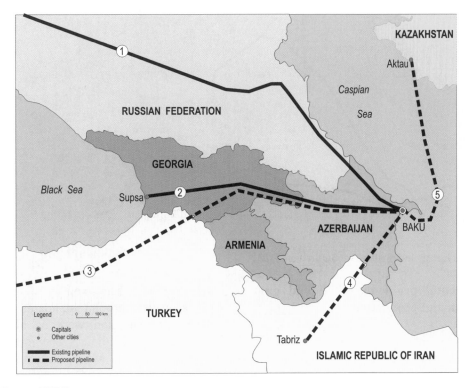

Source: EBRD
Note: 1 – Northern Pipeline; 2 – Western Pipeline; 3 – BTC Pipeline; 4 – Baku-Iran Pipeline;
5 – Trans-Caspian Pipeline

One other proposal that deserves a mention is a Ukrainian pipeline that could transport Azerbaijani oil to Western Europe via Poland. Ukraine has recently built an oil terminal on the Black Sea coast at Pivdenny and, in autumn of 2001, Ukraine (using its own funds) finished building the Odessa-Brody pipeline. There is still a need for a pipeline link from Brody to Plock (in Poland), however. Once that is in place, Azerbaijani oil could be transported westward via the southern Druzbha pipeline or northward to Poland. However, Ukraine has not yet found any customers for the Odessa-Brody pipeline, and Azerbaijan is expressing little immediate interest in the proposal. From Azerbaijan's perspective, there are some technical difficulties and risks involved and possible higher transportation costs. Azerbaijan may further consider this option in the future, should the quantity of oil to be transported exceed the capacity of the BTC pipeline.

Natural gas pipelines

Azerbaijan currently has only limited capacity for the import and export of natural gas. However, efforts are under way to secure export routes and customers for gas deliveries beginning in 2006. (Figure 11.6)

The main conduit for Azerbaijan's natural gas exports will be the South Caucasus pipeline, also known as Baku-Tiblisi-Erzurum, which will run parallel to the Baku-Tiblisi-Ceyhan oil pipeline for most of its route before connecting to the Turkish infrastructure near the town of Horasan. The South Caucasus pipeline is designed for a capacity of 16 billion m^3 a year and is planned to be completed in time for the Shah Deniz project's first contracted exports in 2006.

In 2001, Azerbaijan signed its first major natural gas export deal when it concluded an agreement to begin exporting in 2006 at a rate of 2 billion m^3 a year, increasing to an average 6.3 billion m^3 a year from 2009. Turkey's ability to consume and transit these quantities has been questioned. Greece is also expected to purchase Azerbaijan's natural gas exports. Azerbaijan has already signed a contract with Georgia to export the gas.

Figure 11.6: Proposed pipeline for export of natural Gas

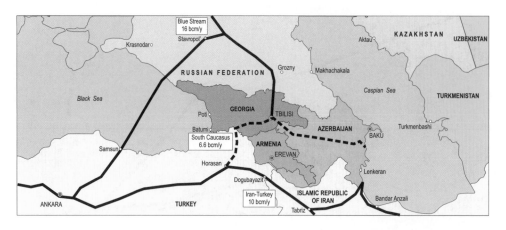

Source: United States Energy Information Administration (EIA).

11.3 Environmental problems in oil and gas exploration and production

The present environmental problems in the oil and gas sector stem in part from past decisions, but they are also the result of a continuing lack of maintenance. The information provided below is derived from material received from SOCAR.

In the initial stage of oil and gas extraction, primitive technology was used. For a long time oilfields were exploited intensively, without the proper technical service available at the time. Production from the wells used open pit, i.e. oil containing contaminated water from the subsoil was refined only to separate oil and water without due regard to the environment. During exploitation and repairs of wells, pressure differentials between the drilling pipe and the drilling mud can cause oil-blowouts, spreading oil, gas and contaminated water. All these events have heavily polluted large areas.

There have been especially high losses to the cultivated lands of the Absheron Peninsula, where about 10,000 hectares of fertile soil used for growing wheat, grapes, figs and other agricultural products have become barren. The oil has penetrated more than 3 m into the soil, and percolation has reduced the quality of the groundwater.

As understanding of the negative environmental effects improved, it became clear that the pollution was even more serious than had been thought. Mining in the area had contributed to pollution by dumping materials with

radionuclides. The intensity of gamma radiation exceeded background standards (15-25 gamma-dose rate/hour) and went as high as 400-1200 gamma-dose rate/hour. The oil refineries also contributed. All waste and residues from the process were dumped at the site. As a result, there has been continuous uncontrolled leakage of large quantities of oil products from non-hermetic reservoirs and underground plants.

Another possible concern is radioactive contamination in residential areas situated in the vicinity of oil fields where surveys have identified contamination by radioactive (low specific activity (LSA)) emissions from the decay of natural radioactive isotopes present in rocks and gases, which concentrates in scales and may be brought to the surface. Experts of the Radiation Medicine Department suspect that an unusually high number of cases of leukaemia or of other diseases in children living in these contaminated areas may be related to exposure to radiation, but no studies have been carried out so far to assess this potential problem.

The standards seen elsewhere in the world for environmental protection during exploration and exploitation of oil and gas offshore were not followed in the Azerbaijani sector of the Caspian Sea. For many years, associated water from the oil reservoirs, sewage, drilling mud and oil containing sand were thrown into the Sea without appropriate treatment. Nothing was done to prevent gushers during exploitation and accidents. The contamination of the Sea from these and the contamination from the pipelines all contributed to the heavy pollution of the sea water with hydrocarbons, heavy metals and other toxic substances from oil and gas extraction.

Exploration and development

The exploration under the new production sharing agreements is performing as well as can be expected. The activities have been strictly monitored, and performance has improved since the pre-1994 activities. Now, before a well can be drilled, an environmental impact assessment (EIA) has to be approved by the Ministry of Ecology and Natural Resources. The EIA deals with the physical environment and with the emissions of the drilling activity. Drilling cuttings may no longer be dumped on the seabed (until recently a common method in many places in the world), nor is it permitted to dump drilling mud into the sea or to discharge associated water (properly treated), not even if the only problem is a salt content higher than that of the Caspian Sea.

As a consequence of the strict regulation, the drilling cuttings are transported onshore, but the high level of drilling mud makes it difficult to deposit. The proposals for onshore use of drilling cuts have been many and innovative. Among the proposals are to use the calorie value of the attached drilling mud in the cement industry by injecting the cuttings in cement kilns.

Oil and gas production facilities

Despite the pre-1994 environmental situation as described above by SOCAR, the absence of intensive investment, of improved maintenance plans and of funds, Azerbaijan states that there are no leaks of substances dangerous to the environment from offshore facilities. The State Control Inspectorate of the Ministry of Ecology and Natural Resources regularly inspects the facilities and during the past 10 years only insignificant leaks have been reported.

The shape of the production facilities and the lack of environmental safeguards are very visible when observing the oilfields in and around Baku. The operational fields have been explored for years and the equipment has not been updated. These fields can produce at low cost only so long as maintenance and capital costs are kept down. If updating the facilities were paid from the revenue, many of the wells would probably be abandoned. Many of the

wells that are no longer operating have been completely abandoned and so have many of the offshore production facilities.

While Azerbaijan is waiting for new offshore facilities to come on-stream, the old offshore and onshore facilities are producing revenue for the State budget. The revenue from SOCAR to the State budget is extremely important for the State's expenditures. Recently the contribution has been more than 60%. SOCAR must therefore also be under extreme pressure, and this might explain the poor-quality of the facilities. However, it would serve the State better if SOCAR were free to evaluate the production facilities and decide either to rehabilitate or to close down facilities that cannot meet environmental standards, including environmental risk exposure.

The fields operated solely by SOCAR are old and close to the end of their life. Both the pressure in the reservoirs and production are decreasing. The level of associated water is increasing, making production costly. Some forecasts predict that the present facilities will stop producing within the next ten years. Whenever the facilities stop producing, Azerbaijan will face large investments to rehabilitate the areas and seal the wells.

Refineries

According to SOCAR, oil and gas extracting and refining are among the main contributors to air pollution in Baku. Despite the decline in extraction and refining, and the series of technical measures taken within the past five years, on average 198,000 tons a year of hazardous matter have been emitted to the atmosphere, including 165,000 tons of hydrocarbons. A substantial part of this hydrocarbon waste is from the surface of the artificial lakes. In addition, although no data or detailed information were available, SOCAR indicated that there was some evidence of radioactivity in the areas around the refineries.

The largest single ongoing rehabilitation effort is in line with a feasibility study on upgrading the refineries.

Domestic refinery capacity adds to the value of exports and keeps the added value in the economy. For the economic development of Azerbaijan the contribution from the refineries could be substantial. However, this requires upgrading the facilities, not only for the sake of the environment, but also to achieve a financially efficient operation of the plant.

However, the BTC pipeline is planned as a crude oil pipeline and the other pipelines out of Azerbaijan are also crude oil pipelines. Unless the export of refinery products becomes possible on a larger scale, the capacity of Azerbaijan's refineries is too large compared to the regional demand for products. Depending on the outcome of the ongoing implementation of the feasibility study, the future of the refineries hinges either on further developing the regional market for products or on constructing a pipeline for the export of their products.

Transport

Production of oil and gas from the ACG and Shah Deniz fields will take place offshore in the Caspian Sea and, apart from terminals at Sangachal (Azerbaijan) and Ceyhan (Turkey), there will be little visible onshore evidence of the projects once construction is completed, since the BTC and SCP pipelines will be buried and the land restored. The terminals will be among the largest of their kind in the world. Construction of each pipeline should take one to two years, involve the transport of large quantities of pipe and other materials, and employ more than 2,000 people at its peak in each country.

The existing pipelines from the offshore production facilities have over time caused some leaks and emissions.

Abandoned offshore exploration and production sites

The environmental problems from abandoned production sites are all connected to the pre-1994 production facilities. In permits given around the world, terms for production also include terms for abandoning sites. Among these are the removal of production facilities, the sealing of wells, and the cutting of well heads a number of metres below the seabed or the surface. These terms have not been imposed in Azerbaijan.

In the Azerbaijani part of the Caspian Sea several production facilities have been abandoned,

leaving all facilities intact, subject only to erosion to finally clear the area. The wells are left open, the platform is left on the site with most of the substances that could potentially contaminate the sea, and the pipes are left on the seabed.

The present offshore activities developed prior to 1994 and their environmental impact result from four sources: leaks from wells, substances used in production on platforms, the platforms themselves and connecting pipelines.

Leaks from wells

Normal practice would be to seal up a well and cut the string at least 10 metres below the seabed, leaving the seabed intact. However, in Azerbaijan, most of the wells, offshore as well as onshore, have been left open when abandoned. The risks for the environment are many. In 2001, an inspection of 200 abandoned offshore wells and 50 to 60 abandoned platforms indicated that there was no imminent threat to the environment, but the descriptions of some of the wells left open the possibility of future problems. Furthermore, there are many more abandoned wells than the ones included in the inspection.

Substances used in the production on the platform

The international norm would be to remove all substances that could potentially damage the environment. Unfortunately, there was no information available on the current status of abandoned platforms, but it would be difficult to imagine that they had been properly decommissioned. While some facilities might never have used harmful substances during operation, others might have used hydraulic oil, cutting oil, and other petroleum products and chemicals.

Platforms

If the wells are properly sealed and the platform properly cleared of all materials and substances that could damage the environment, only visual pollution is left. If the platforms become a danger for competing uses of the sea, their presence should be reconsidered. The actual situation of the abandoned platforms in Azerbaijan is not known.

Pipelines

Like the other production facilities, the pipelines were left where they were once used. The pipes might be completely empty, but they might also be full of seawater, mixed with associated water and some oil. Only inspection of the pipes can determine their present danger to the environment.

Abandoned   onshore   exploration   and production sites

When facilities and fields were abandoned, it appears that the only actions that were taken were to stop the machinery and, if appropriate, to realize the value of the facilities. Oil production has been going on for so many years that registration of abandoned wells might not even be possible. At the same time, ongoing production continues to contaminate some of the sites that are likely to be abandoned in the future.

Many of the fields that have been abandoned were producing from reservoirs close to the surface. The reservoirs are evidence that there are subsoil sealings, but the migration of hydrocarbons not trapped by a sealing will migrate further upward. In a country that rich in oil deposits, there is significant concern about possible oil in the soil, not only from oil production but also from natural contamination. However, this should not be used as an excuse for depositing substances in open ponds and without proper sealing. The open depots and the open wells allow leaks from one environment to another. This might not be severe at present, but, as migration in the abandoned reservoirs fills the structure, it could become so.

11.4   Policy objectives and management

Policy framework

The environmental policy of Azerbaijan for the oil and gas activities is clear in the sense that no further environmental damage from them should occur. The policy is reflected in the production sharing agreements, which also require an environmental impact assessment before new activities can begin. This should mean that all future production-related activities will be environmentally benign. It does not, however, address the environmental load from past production and present production in old facilities, which is not up to the standards of the PSAs.

It is a declared objective of the production sharing agreements to bring in foreign companies that are able to develop oil and gas production facilities that meet international standards. However, specification of the term "international standards" is not precise and could become synonymous with best available technology (BAT). BAT may be attractive, but it could also exceed the standards necessary for Azerbaijan, raising the cost unnecessarily.

The State Control Inspectorate of the Ministry of Ecology and Natural Resources makes unannounced inspections of oil installations – though foreign companies require 24 hours' warning. The Inspectorate has access to transport and equipment for inspection. The most common speedy transport in the offshore industry is a helicopter. Inspectors may use these helicopters, but only upon request to the oil companies, making "surprise" visits impossible.

Legal framework

There is no specific law for the oil and gas sector. There are two laws that regulate this field of activities: the Law on Energy and the Law on Use of Energy Resources. A draft oil and gas law was developed and discussed in parliament in 1997. Both the first and the second drafts were rejected. By mid-2003 there was no indication of when or if a new draft law would be prepared for parliamentary considerations.

However, the subsoil resources are the property of the State. In accordance with the Constitution and the constitutional Act of State Independence and the Law on Subsoil, this includes ownership of all petroleum existing in its natural state underground or subsurface, including the portion of the Caspian Sea within Azerbaijan's jurisdiction.

The legislative power rests within Parliament and administration with the Ministry of Fuel and Energy. The Ministry nominally has the power to contract companies to manage the exploration and exploitation of oil and gas on behalf of the State, when the draft contracts have been considered by

Parliament. The State Oil Company participates along with private companies in the exploration and extraction activities.

Economic instruments

Royalty, or mineral fields tax, is part of the general tax regime and can range from 3% to 26%. Royalty on natural gas was 20%, and on oil 26% in 2000. The royalty is deductible for profit tax purposes. The revenue that comes from taxation is a very important part of the State budget, providing up to 60% of the total budget. The payments to the State from SOCAR and PSAs are collected in the State Oil Fund.

The State Oil Fund (SOFAZ) was set up in accordance with a presidential decree (1999). It is a mechanism whereby energy-related windfalls are to be accumulated. The objective of the Fund is the professional management of oil- and gas-related revenue for the benefit of the country and its future generations. Its statutory regulations were approved by presidential decree in December 2000.

The major sources of SOFAZ income are:
- Revenues generated from the country's share of sales of crude oil and gas (after deductions stipulated in the appropriate legislation);
- Bonus payments, depending on events like first oil discovery;
- Royalties (acreage payments);
- Rents for the use of State property by foreign companies under oil and gas contracts;
- Revenues generated from the sale of assets that are transferred to Azerbaijani ownership under contracts signed with foreign companies;
- Other revenues from joint activities with foreign companies;
- Revenues generated from investment of the Fund's assets.

SOFAZ will provide a mechanism for the controlled funding of economic development, especially major infrastructure projects. Furthermore, SOFAZ will play a major role in ensuring that there are safeguards in the country's economic management system at a time when Azerbaijan has an economy in transition and a vast economic and social development agenda.

In practice the State Oil Fund is controlled by Azerbaijan's President, and so control over the revenue from the oil and gas activities is divided between the President and Parliament, which controls the revenue from taxation and SOCAR income.

Institutional framework

The subsoil and the oil and gas reserves are the property of the State. The State contracts companies to exploit the reserves on its behalf. The interest of the State is represented by the Ministry of Fuel and Energy.

SOCAR was established in September 1992 with the merger of State Oil concern Azerineft and Azneftkimiya Production Association. SOCAR and its many subsidiaries are responsible for producing oil and natural gas in Azerbaijan, operating the country's two refineries, running its pipeline system, and for managing its oil and gas imports and exports. SOCAR is party to all the international consortia developing new oil and gas projects in Azerbaijan. The company employs 60,000 people. In January 2003, Azerbaijan's President issued a decree calling for the reorganization of SOCAR. According to the SOCAR Chairman, the reorganization should be complete by the end of 2003.

The Ministry of Fuel and Energy was established in April 2001 and, inter alia, is responsible for oil and gas activities. The Ministry is, in fact, supposed to represent the owner, i.e. the State, in its commercial activities within the oil and gas sector. This should entail responsibility for contracting companies, including SOCAR, for exploration and extraction. However, until the Ministry is fully staffed, SOCAR continues to negotiate and manage contracts. Among other things, this means that SOCAR is both operator of offshore activities, partner in the 22 new contracts and, at the same time, the administrative unit tendering and issuing production sharing agreements. There is virtually no transparency in such an arrangement.

The Ministry of Fuel and Energy monitors SOCAR and signs the contracts.

11.5   Conclusions and Recommendations

Conclusions may be divided into the following issues: (1) concerns related to the PSAs; (2) concerns related to pre-1994 development and producing facilities; (3) abandoned facilities; (4) refineries; and (5) access to information.

So far, there has been only limited experience with PSAs, and it is therefore impossible to evaluate them at this stage. The regulation concerning PSAs seems to form a thorough basis for protecting environmental performance. Furthermore, there is both the institutional capacity and a willingness to use the available means and instruments.

The offshore hydrotechnic facilities are apparently in their later stages of production and, for many, decommissioning will need to be considered within a few years. The inspections of pre-1994 facilities do not seem convincing. The Inspectorate has reported only "insignificant leaks" within the past 10 years, yet this concerns facilities that have openly been acknowledged as heavy polluters throughout the decade. In addition, this improvement was supposedly achieved without comprehensive investments or an improved maintenance scheme.

If the facilities are meeting international environmental standards, as the Inspectorate claims, pollution can only have come from abandoned facilities, mismanagement or poor operation of facilities. This is unlikely, and therefore some facts not available to the EPR mission might justify a revision of the conclusion. It is more probable that pollution from the older facilities continues much the same. The only difference may be a reduction in performance due to a lack of maintenance.

The State Control Inspectorate for Environment and Use of Natural Resources under the Ministry of Ecology and Natural Resources has inspected the abandoned facilities, and reported that there was no evidence of threats to the environment. However, the number of abandoned wells in the Caspian Sea is much higher than the 200 inspected. Information provided is that wells have been abandoned according to the world standards, and no leaks have been registered for the last 50 years.

The method used for sealing wells is not conclusive, since the explanation varies according to which State authority provides the information. What can be concluded is that only part of the wells is filled with material, but which part is unclear. The problems of the refineries and the future need for capacity will be included in the ongoing feasibility study. Should the study conclude that it is worthwhile to continue

operations, the contamination of the site and the technology used might be dealt with. The worst case is if the feasibility study concludes that it is not financially feasible to continue operations after investments and cleaning the site to an appropriate level.

SOCAR was not particularly open to providing information on pre-1994 production facilities. Given the importance of transparency and that Azerbaijan has ratified the Aarhus Convention, more attention should be given to providing good information.

It is important that Azerbaijan prioritize its efforts to improve the environment. With the new PSAs, the State has taken care of the future efforts in the oil and gas sector. However, it is essential that attention also be given both to stopping pollution from present production and, once that is under control, to cleaning up the contaminated sites. The level of clean-up should be based on feasible standards.

Registering the contaminated sites is a huge task, which could involve the 28 regional departments for environment and natural resources. Also, determining the desired level of clean-up should include the regional levels. The financial means for this work could come from several sources. The Oil Fund should be an obvious source as its objective is to save for future generations.

Recommendation 11.1:
The Ministry of Ecology and Natural Resources should:
(a) Register contaminated sites and identify the level of contamination;
(b) Determine the methods used for sealing offshore wells;
(c) Determine the standards for the clean-up of contaminated offshore and onshore sites; and
(d) Provide financial means for the work to be undertaken to seal the wells, either from the State budget, the Oil Fund, or an environmental royalty on present production.

Recommendation 11.2:
(a) SOCAR should begin the transition to divesting itself of the responsibility of negotiating and approving contracts with foreign companies so that it may concentrate on its managerial responsibilities and implement fully the presidential decree of January 2003 calling for its reorganization. and

(b) *The Ministry of Fuel and Energy should complete its staffing and strengthen its capacity to be able, at the earliest possible opportunity, to assume all its legal responsibilities, including that of negotiating contracts.*

Oil and gas activities have huge impact on the environment. To provide an opportunity for all stakeholders to participate in decisions that may affect their economy, society and environment, it would be useful to establish advisory boards for offshore and onshore activities. The boards should include representatives from all relevant ministries, local economic interests and non-governmental organizations.

The boards should act as an oversight group for the oil and gas industry, and establish a dialogue between the oil companies and other stakeholders in the country. This could include, for example, establishing clear guidelines regarding the information, and its timeliness, that the oil companies, including SOCAR, would be obliged to share with the public; full transparency regarding PSAs; full transparency regarding the State Oil Fund; the establishment of a system for receiving and responding to complaints from individuals and organizations, and other matters considered important to the stakeholder. If the matters at hand so warrant it, members of the boards could have the option of inviting independent experts to investigate an issue.

Recommendation 11.3:
The Cabinet of Ministers should establish two advisory boards -- one for offshore and one for onshore activities, each with representatives from relevant ministries, including, in particular, the Ministry of Fuel and Energy and the Ministry of Ecology and Natural Resources, local economic interests and non-governmental organizations.

The advisory boards should be supported by a secretariat able to call in independent investigations. The boards could also play a major role in the work for recommendation 11.1.

Experience with PSAs to date is too limited to be able to evaluate them effectively. While the contracts themselves take into account all international norms and standards for oil and gas exploration, it is impossible at this early stage to assess their implementation. The first PSA was signed nine years ago, but more than half have been signed only within the past five years, and 87% in the past six years.

Recommendation 11.4:
The Ministry of Ecology and Natural Resources, in cooperation with the Ministry of Fuel and Energy, should assess the environmental impact of the activities being undertaken under each production sharing agreement within five years after the start of operations, at regular intervals thereafter and after a site has been abandoned.

Chapter 12

HUMAN HEALTH AND ENVIRONMENT

12.1 Overall health status and environmental conditions

Demographic trends and population characteristics

In 2002 the total population of Azerbaijan stood at 8,141,400, of whom 50.7% lived in urban areas and 49% were male. Nearly 30% of the population was younger than 14, while people aged 65 and above comprised 6.3% of the total.

The annual population growth rate between 1975 and 2001 averaged 1.4%, and is projected to be 1.0% from 2001 to 2015. While the fertility rate for women for 2000-2005 is estimated at 2.1%, less than half of what it was during 1970-1975 (4.3%), the total population is expected to grow to as much as 9.5 million by 2015. Urbanization is expected to continue, and the estimated urban population in 2015 is expected to be 53.9% of the total.

Overall, people's health status has improved since the mid-1990s. For some indicators, trends and levels are similar to those seen in neighbouring Caucasian or Central Asian countries.

Life expectancy suffered a sharp decline over the 1992-1994 period. To a significant extent, this was because of the armed conflict with Armenia, accompanied by the socio-economic difficulties experienced in the transition period. Life expectancy began to rise from 1995. In 2001, it was 72.4 years, in line with that of other Caucasian countries and higher than that of the Commonwealth of Independent States (CIS) (67.1 years) (Figure 12.1).

Persistent pollution from Soviet-era industrial activities, such as those carried out in Sumgayit, and long-term occupational exposure are also significant factors contributing to the burden of disease and disability. These are particularly acute in large economic centres such as Baku, Sumgayit (see box 12.1), Ganja and Ali-Bayramli.

Figure 12.1: Life expectancy at birth, 1990-2001

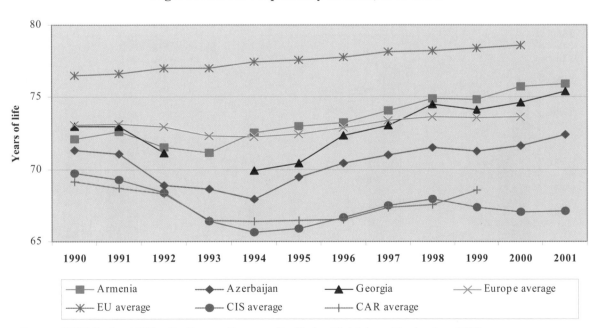

Source: WHO Regional Office for Europe. European Health for All database. Version June 2003.

The high prevalence of poverty and the relatively large proportion of refugees and internally displaced persons (IDPs) in the population are also important socio-economic health determinants that should not be underestimated. The Household Budget Survey of 2001 estimated the country's poverty at 49% of the total population and extreme poverty at 17%. Poor families, having fewer choices and opportunities, compromise their health needs and engage in risky behaviour. This includes an inability or unwillingness to seek costly medical services, as well as a reliance on self-treatment, traditional service providers and birth attendants (midwives). Lack of transport services to the clinics prevents the population in remote areas from benefiting from their services.

Refugees and IDPs created by the Nagorny-Karabakh conflict live in inappropriate conditions that are harmful to their own health as well as to the environment in which they live. The number of refugees and IDPs at the beginning of 2002 was estimated to be 211,200 and 572,000, respectively, constituting 9.6% of the total population of Azerbaijan. The problems associated with their living environment, as well as their social and employment status, contribute considerably to Azerbaijan's overall poverty and environmental degradation. Poverty among IDP households stands at 63% and their health indicators are often below the national average.

Mortality trends

Official data show that the standardized mortality rate for all causes of disease and injury in Azerbaijan was slightly higher than the average for the WHO European region in 2000. It has, however, been declining, with a peak of 1,146.09 deaths per 100,000 population in 1994, and a low of 989.02 per 100,000 in 2001.

According to the data, diseases of the circulatory system are by far the most widespread cause of death (609.89 per 100,000 population in 2001). Premature (i.e. 0-64 years) mortality from cardiovascular diseases is similar to the CIS average, but significantly higher than the West European average.

Malignant neoplasms are the second most frequent underlying cause of death. Mortality from malignant neoplasms in Azerbaijan has long been lower than the average rates of the overall European region, Central Asia and the EU, and is similar to those of Georgia and Armenia. The data do show a declining trend since the late 1980s, from a peak of 148.36 per 100,000 down to 104.78 in 1998. The official death rate from cancer in 2001 was 110.51 per 100,000 population. (Figure 12.2)

Figure 12.2: Standardized mortality rates by selected causes of death, 2000

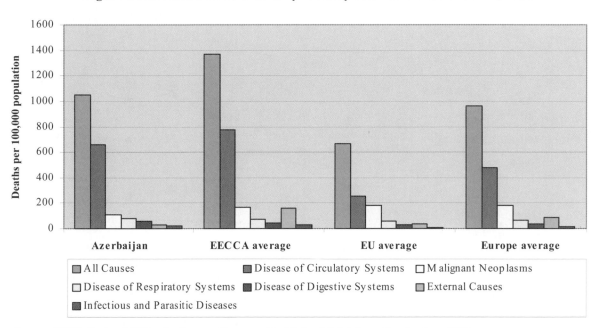

Source : WHO Regional Office for Europe. European Health for All database. Version June 2003.

Figure 12.3: Morbidity rates for selected causes of disease per 100,000 population, 1996-2000

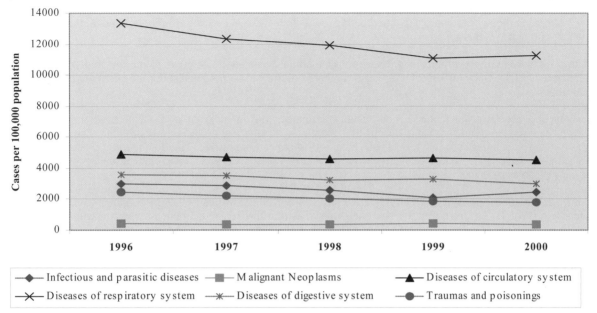

Source: WHO Regional Office for Europe. European Health for All database. Version June 2003.

On the other hand, at 79.38 per 100,000 population in 2000, mortality from diseases of the respiratory system is higher than the average rates for CIS (74.65), Europe (62.94) and the EU (58.37). In Azerbaijan it peaked at 113.47 per 100,000 population in 1994 and by 2001 it had gone down to 65.63.

Mortality from diseases of the digestive system has increased since the late 1980s, and for a number of years it has been close to the Central Asian average, remaining considerably higher than the CIS average (56.56 per 100,000 population against 48.35).

The trend in mortality from external causes of injury and poisoning is substantially different in Azerbaijan than in most other countries in Eastern Europe, the Caucasus and Central Asia (EECCA), above all owing to the peak in such deaths in 1992-1994 caused by the armed conflict in Nagorny-Karabakh. This is apparent from the number of deaths coded as homicide during this period. Lately, this indicator is among the lowest in the European region.

Traffic-related mortality is one of the lowest in the European region (5.7 per 100,000 population in 2000, i.e. nearly half the EU average) with approximately 550 deaths a year. However, the very sharp decline in mortality rates reported between 1993 (13.7/100,000) and 1994

(2.27/100,000) in the absence of major policy initiatives in this area seems to reflect the decline in transport activities for both goods and passengers observed in the mid-1990s rather than the result of comprehensive road safety policies.

In Azerbaijan, as in Central Asia, the mortality rate from infectious and parasitic diseases has for a long time been above the EECCA average. However, the situation was reversed in 2000 and 2001, when the rates for Azerbaijan fell to 23.29 and 22.8 against EECCA averages of 26.22 and 25.51 per 100,000 population, respectively. (figure 12.2)

Although the rate is declining, childhood (under 5 years of age) mortality from diarrhoeal diseases is still relatively high at 31.18 per 100,000 population in 2001 in comparison with the EECCA average of 21.86. The EU average in 2000 was 0.49. The maternal mortality rate in Azerbaijan remains high, the main cause being haemorrhage during delivery.

Morbidity trends for environment-related diseases

According to the Ministry of Health, total morbidity declined from 38,628 per 100,000 in 1996 to 33,434 per 100,000 in 2000 (Figure 12.3). While the diseases contributing most to the population's mortality are of the circulatory system, by far the largest factor of morbidity is diseases of the respiratory system.

Box 12.1:   Cancer incidence in the industrial city of Sumgayit: a descriptive study

The available data suggest that, in the early 1990s, cancer rates in Sumgayit were higher than in the rest of the country, similar to those in Armenia and Georgia, but lower than those in the Russian Federation, and only one third to one half of those in Canada.

Results of a lifestyle survey conducted to detect possible confounders suggest that smoking, diet and drinking habits are similar among the study regions of Azerbaijan, and are not likely to be responsible for the observed differences in cancer rates.

The study suggests that the higher cancer rates in Sumgayit are most likely a result of exposure in industry. However, the extent of the difference is difficult to estimate because of inadequacies with the data. The cancers that are the highest in Sumgayit are urinary bladder, lung and all cancers combined. The documented presence and use in the Sumgayit factories of a number of chemicals known to be or suspected of being carcinogenic in humans lends plausibility to the study's findings.

The decrease in cancer rates observed during the early 1990s in Sumgayit, as well as throughout Azerbaijan, and in Armenia and Georgia, suggests that the decrease was related to the social, economic and administrative changes that followed the end of the Soviet Union. Owing to the difficulties that these countries faced at independence, cancer diagnosis and cancer reporting likely dropped, even though the actual number of cancers may not have changed. Cancer under-reporting is also likely to explain the lower than expected differences in cancer rates in Azerbaijan relative to Canada.

In 2001, the morbidity rate for diseases of the respiratory system was 11,274 per 100,000, and for diseases of the circulatory systems 4,516 per 100,000. Diseases of the digestive system, infectious and parasitic diseases, traumas and poisonings, and malignant neoplasms follow with rates of 2,964, 2,446, 1,830, and 349, respectively, per 100,000 population. While they all declined over the 1996-2000 period, the improvements are slow, except for diseases of the respiratory tract, which had declined to almost the EECCA average by 1999. (See figure 12.3.)

The incidence of malignant neoplasms (i.e. the registration of new cases of cancer per year) appears to be abnormally low (67.7 per 100,000 population in 2001 compared to averages of 257 per 100,000 in EECCA and 446 per 100,000 in the EU), pointing to the possibility of very serious under-reporting of new cases, as observed in a recent study on cancer. (See box 12.1.)

Poliomyelitis was eradicated in the mid-1990s and nearly 100% of infants are reported to be immunized against it. Incidence rates for acute intestinal infection of known and unknown aetiology, salmonellosis, brucellosis, bacillary dysentery and typhoid fever remained relatively constant between 1996 and 2002. (See figures 12.6 and 12.7.) While reducing the rates of these diseases remains a challenge, a major effort is needed to raise the level of detection of acute intestinal infections.

Box 12.2:   Tuberculosis (TB)

In 1990, 2,506 new cases of TB were registered, and the number increased to 5,113 (64 per 100,000 population) in 2000. The number of new registered TB cases went down to 4,877 (60 per 100,000) in 2001 and 4,422 (54 per 100,000) in 2002. The mortality rate from TB also decreased from 6.2 per 100,000 population in 2000 to 3.7 per 100,000 in 2002.

Compared to the CIS average in 2001, where new registered TB cases per 100,000 population was 88, the situation in Azerbaijan may not appear too critical. However, it is significantly higher than the EU level of about 11, or the East European average of about 50 per 100,000 (see figure 12.5). While there is no specific evidence, it is likely that most of these people are poor, since TB spreads more rapidly in poor sanitary and housing conditions (important aspects of environmental health), overcrowding, deteriorating socio-economic conditions and inadequate treatment.

The Ministry of Health, jointly with WHO, has promoted the "directly observed treatment" (DOT) strategy, which was implemented in 1995 in three pilot districts. This programme focuses primarily on training doctors and nurses to treat patients with free medications taken under direct supervision. It also includes educational programmes for both TB and healthy lifestyles, as TB is best tackled through timely intervention and is easier and cheaper to cure in the early stages. The Government will also cooperate with international donors to improve the population's BCG vaccination coverage against the disease.

Figure 12.4: Viral hepatitis incidence rates per 100,000 population, 1991-2001

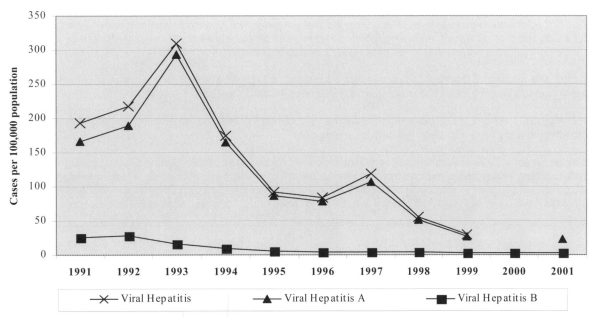

Source: WHO Regional Office for Europe. European Health for All database. Version June 2003.

Incidence rates of viral hepatitis have decreased dramatically – from a high of 310 per 100,000 in 1993 to 30 per 100,000 in 1999 – owing mostly to a reduction in the rate of hepatitis A, although the possibility of under-reporting may not be ruled out.

While malaria and tuberculosis rates have also declined drastically, they continue to remain the country's major health concerns. (See boxes 12.2 and 12.3 and figure 12.7.)

There is a suspicion that the number of cases of diseases is under-reported, partly because many of them are diagnosed by private facilities and partly because in many cases no doctor is consulted nor is there any laboratory confirmation of the diagnosis.

12.2 Environmental conditions associated with health risks

Health effects of air pollution

The link between exposure to air pollution and a broad range of health effects is well-established. These include: total mortality and morbidity from respiratory and cardiovascular diseases, the exacerbation of asthma attacks and an increase in their frequency, and neuro-developmental disorders related to children's exposure to lead. However, the lack of monitoring data for the appropriate indicators (e.g. particulate matter), of information on population exposure and of studies investigating the link between air pollution and health in the population, makes it impossible to produce

quantitative estimates of the disease burden attributable to air pollution in the country. On the basis of epidemiological evidence available from other countries, it is fair to assume that also in Azerbaijan, and especially in its largest cities, there is a significant burden of morbidity and mortality associated with exposure to air pollutants emitted from stationary and mobile sources.

The Ministry of Ecology and Natural Resources monitors all industrial activities and facilities to determine their overall pollution impact. The Ministry of Health is responsible for workplace health and safety. The two do, however, work closely together in developing hygiene standards and norms. (See chapter 5 on air management and transport.)

There is interest in having the Ministry of Ecology and Natural Resources work with the Ministry of Health to examine the effects of air quality on human health. This is deemed possible because these Ministries have worked together in the past. Research into air quality, and noise and health has been conducted and published locally, including in local scientific journals.

Very little information is available on indoor air pollution. The relatively high prevalence of smoking (26.5% of people older than 15 are estimated to be regular daily smokers) and the continued use of asbestos as a construction material represent the main sources of potential exposures to well-known carcinogens.

Box 12.3:   Malaria

Eighty per cent of Azerbaijan is considered a malaria-endemic zone. The most recent malaria epidemic was in 1994, when 667 infections were registered. It is suspected that stagnant river flow and swamps created from the rising level of the Caspian Sea were the contributing factors. The number of total cases continued to grow and peaked at 13,135 in 1996. The national programme to roll back malaria is expected to continue until 2004.

In 1999, the number of reported cases was less than half that in 1998 at 2,315. As a result of comprehensive anti-epidemic measures, the incidence came down to 506 cases in 2002.

As of May 2003, 67 cases had been reported for the year. While the infected population has been treated, the natural hotbeds of malaria have also been reduced in recent years.

However, in early 2003, several major floods took place in different parts of the country, creating conditions favourable to the breeding of anopheles mosquitoes, and raising the risk of a new wave of malaria. Floods are suspected to be a result of climate change – heavy rain and increased water levels of the Kura and other rivers. While some large-scale outbreaks have been expected, the country is not able to make projections on the extent of the consequences.

Water, sanitation and health

The most important water-related diseases are dysentery, typhoid fever, hepatitis A, and acute intestinal infections of known and unknown aetiology (Figures 12.6 and 12.7). The National Environmental Action Plan (NEAP) identified the deterioration of water quality in both rural and urban areas, and the related increase in water-borne diseases, as one of the country's main environmental problems. The State Programme on Poverty Reduction and Economic Development for 2003-2005 also recognizes that one of the primary causes of morbidity and mortality in children is diarrhoeal disease, usually resulting from contaminated water, which contributes to aggravating malnutrition by reducing the capacity to absorb the few nutrients available.

Although, according to official statistics, the number of cases of water-related diseases such as dysentery and typhoid fever have generally diminished in recent years, these numbers may not reflect reality, particularly among the poor and the very poor. The information from the Ministry of Health suggests that 9% to 13% of families have reported diseases caused by water. Groups of families who have daily access to water have a very low water-related disease burden. However, clean water is becoming a privilege of the rich; they can afford filters and bottled water.

Figure 12.5: Tuberculosis incidence rates per 100,000 population, 2001

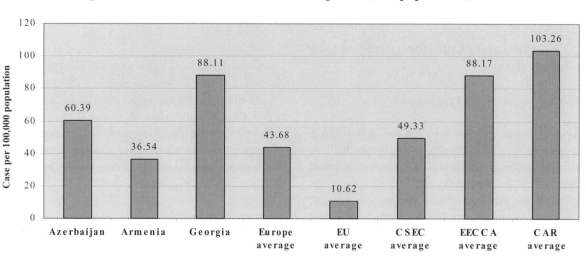

Source: WHO Regional Office for Europe. European Health for All database. Version June 2003.

Figure 12.6: Water-related and food-borne infection rates (all ages), 1997-2002

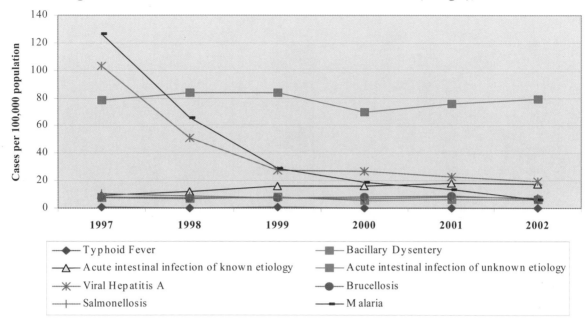

Source: Ministry of Health. Infectious diseases.

Providing water to the population is the responsibility of local and central authorities, depending on the geographic location. For example, in the Absheron Peninsula (where the two largest cities are located) the responsible body is the Absheron Regional Joint-stock Water Company, which has the authority of a ministry. National data on water supply and sanitation coverage are not reliable enough to make them the basis for policy-making and intervention. The Multi-indicators Cluster Survey (MICS) of 2000, nevertheless, shows some trend in water supply and sanitation. According to the survey, 76.3% of the population uses drinking water from improved sources, with some significant regional differences and differences across socio-economic strata. However, this may still be an overestimate because it assumes that all piped water is safe.

Even where there is piped water, its quality is not always reliable. The Ministry of Health is responsible for setting the quality standards for drinking water, monitoring the water quality after treatment and in the distribution network, through its local sanitary and epidemiological centres, and for monitoring the quality of sewage discharged into rivers. However, the lack of adequate equipment and overly strict standards prevent it from carrying out its functions adequately. The Government intends to establish a system of regular and reliable monitoring of water supplies.

While national statistics do not make it possible to pinpoint areas with particular water quality problems, monitoring at the national level indicates that between 1996 and 2002 approximately 20-28% of water samples did not meet chemical standards, and approximately 13-17% did not meet bacteriological requirements. Reconstruction of water purification stations, as a part of the greater Baku water supply rehabilitation project is intended to improve these poor results.

MICS shows that 80.8% of the population lives in households with sanitary means of human waste disposal. This percentage peaks for rich households (100%) and households in the Baku area (99%). Rural sanitation is considered a private household issue, and few have access to even basic sanitation. The only official government involvement is through the State Sanitary Epidemiological Service of the Ministry of Health, which advises and approves the location of latrines through the regional centres.

The sewage system in Baku serves about 78% of its population, including IDPs and refugees. At present, only 50% of the sewage is treated in greater Baku. The remaining part is discharged without treatment into the Caspian Sea. In other urban regions, the provision of sewage drops to about 32%.

Figure 12.7: Water-related and food-borne infection rates, 2002

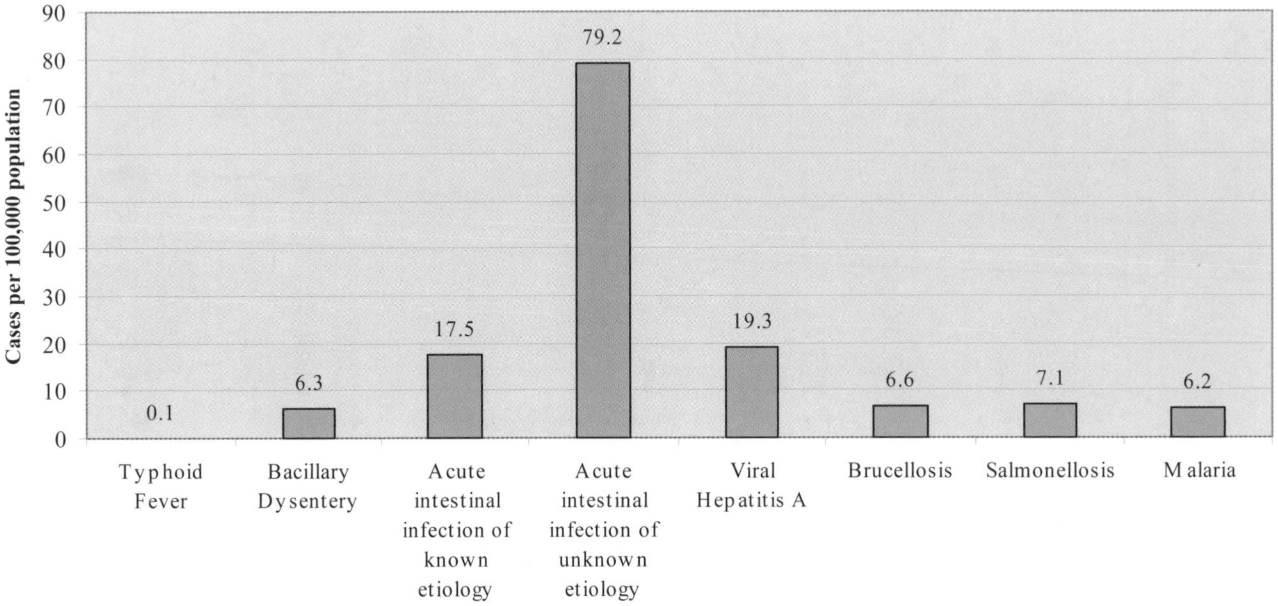

Source: Ministry of Health. Infectious diseases.

The floods of early 2003 are expected to add to drinking water contamination in the rural areas and, thus, to water-borne diseases. The situation further increases the threat from the rising level of the Caspian Sea, which not only brings the contaminated water closer to the land, but also does not accommodate drainage for river waters, which are also polluted with industrial and human waste. In addition to efforts to pump water out following natural disasters, special teams were organized to visit the affected areas to identify populations with intestinal infections among those who provided specimens for analyses. The teams also implemented prophylactic measures including advisories to boil water. Televised messages to the population were also sent, inviting people to report to doctors if they had symptoms.

Food safety and nutrition

Both domestically produced and imported foods are monitored by analysing randomly selected samples, and a certificate of quality is issued to those complying with the standards. Domestically produced food is analysed at laboratories of the State Sanitary and Epidemiological Service, and imported food is analysed at laboratories of the Ministry of Trade.

The Law on Food Products (18 November 1999) applies to imported as well as domestic food. However, owing to the laboratories' limited capacity, food products from countries where there is more stringent quality control will receive a certificate of quality without laboratory analysis in Azerbaijan.

Table 12.1: Monitoring of Drinking Water in Azerbaijan, 1996-2002

	1996	1997	1998	1999	2000	2001	2002
	Sanitary-bacteriological examinations						
Number of total samples	22,242	22,525	20,254	20,045	22,017	23,189	23,211
Of which those did not comply with the standards	3,672	3,516	2,898	2,784	2,820	3,160	3,211
Percentage of non-compliance	16.5	15.6	14.3	13.9	12.8	13.6	13.8
	Sanitary-chemical examinations						
Number of total samples	24,401	22,289	21,956	21,751	24,090	24,686	24,543
Of which those did not comply with the standards	6,755	5,178	4,945	4,625	4,883	5,485	5,341
Percentage of non-compliance	27.7	23.2	22.5	21.3	20.3	22.2	21.8

Source: Ministry of Health, 2003.

The degradation of land, leading to soil erosion and salinization, is affecting the ability of the rural population to use the land to generate income. Pollution from fertilizers and pesticides and the lack of effective controls on the food chain increase the risk of human exposure to these harmful substances through the consumption of contaminated food. However, no data seem to be available to estimate the extent and seriousness of this potential threat.

As part of the State Programme on Poverty Reduction and Economic Development for 2003-2005, Azerbaijan will improve the coverage of veterinary vaccinations and introduce epizootic control measures to improve the quality of livestock and reduce the risk of zoonotic brucellosis and tuberculosis being transmitted to humans. It is also important to improve the control of food safety by the State laboratories and to upgrade quarantine control at international borders.

Among the most serious infectious diseases caused by microbiological contamination of food reported to the WHO Surveillance Programme for Control of Food-borne Infections and Intoxications in Europe are:

- Reported cases of salmonellosis in the country have declined, from 2,415 cases in 1993 to 686 cases in 1998 and 551 cases in 2000; however, under-reporting may not be ruled out.

- During 1993-2000, 34 food-borne disease outbreaks were reported to the WHO Surveillance Programme for Control of Food-borne Diseases in Europe, of which 18 were due to Cl. botulinum, with 295 cases in total. The most frequently involved food in these investigated botulism outbreaks include home-canned vegetables and meats and smoked fish and sausages, which are commonly contaminated by inadequate control of the temperature during canning and by food from unsafe sources. This points to the need for better educating consumers in good hygiene and correct home canning.

Interruptions in the cold chain and a lack of adequate and constant refrigeration are important risk factors for the microbiological contamination of food during storage and transport. Dairy products, including those for infants, are particularly at risk, and contamination appears to increase during summer and in areas where water supply and sanitation systems are poor.

Despite recent improvements in nutritional status connected with positive developments in agriculture and nutrient production, the average diet still lacks adequate consumption of all major calorific sources. Twenty-three per cent of the population is estimated to be undernourished, while 17% and 20% of the population under the age of 5 is underweight and underheight, respectively. Ten per cent of infants are born with low birth weight. The situation of refugees and IDPs is even more dire.

Hospital waste

Both in urban and in rural areas, hospital waste is disposed of together with municipal waste, after disinfecting contaminated material by soaking them in hypochlorite solution. Syringes and sharp materials are disposed of in separate rigid containers, which, at least in the largest hospitals, are collected by the chief nurse for separate disposal.

The International Medical Corps is implementing primary health care programmes in six rural regions in the south of Azerbaijan (Saatli, Sabirabad, Imishili, Bilasuvar, Beylegan and Fizuli), which installed incinerators in the health facilities of the communities serviced by the programmes. The design of the units has been approved by the State Sanitary Epidemiology Service. (See also chapter 6, management of waste and contaminated sites.)

An important disease related to the safe disposal of infectious hospital waste, and in particular the safe handling of blood and body fluid samples, and the disposal of sharp items, is hepatitis B. According to expert judgment, official statistics for hepatitis B are likely to severely underestimate the magnitude of the issue (perhaps by a factor of 5 to 10), as the number of cases diagnosed by private laboratory facilities are not notified to the Ministry of Health. The vast majority of cases of hepatitis B occur among medical professionals. Injection drug use does not seem to be a significant problem. Vaccination programmes for hepatitis B are virtually non-existent, although a priority group has been identified among children born to mothers carrying the virus.

Another potentially underestimated problem is related to hepatitis C, for which no screening in blood or blood products is available. This is a potential problem for thalassaemic patients, who are particularly at risk because of exposure through periodic transfusions.

Ionizing radiation

The main sources of potential exposure to ionizing radiation for the general population and occupationally exposed workers are diagnostic activities (X-rays), and radioactive materials used by medical and research institutions, as well as in the oil and chemical industry and the military sector.

The geological nature of materials used in construction could lead to indoor exposure to natural radon gas, but there is no equipment to assess this potential source of exposure for the general population. Another possible concern is radioactive contamination in residential areas situated in the vicinity of oil fields, as this could, at least in theory, lead to an excessive number of cases of leukaemia or of other diseases related to exposure to radiation in children living in these contaminated areas. However, no studies have been carried out so far to assess this potential concern (see chapter 11 on environmental concerns in the oil and gas sectors).

Radioactive material can be imported with permission from the State Control Inspectorate and written permission from the Prime Minister. At the moment, 482 facilities are mapped as sources of radioactive waste, and information on the type and quantity of isotopes used is available from the State Control Inspectorate and the Radiation Medicine Department of the State Sanitary Epidemiological Service. The Radiation Medicine Department is also involved in testing radioactive contamination of construction material and food products (though on a selective basis rather than on the basis of a regular monitoring programme), and in the control of occupational exposure. It also serves as a focal point for the international alert system of WHO.

International collaboration with the International Atomic Energy Agency (IAEA) on training and technical assistance (e.g. in the provision of equipment such as computers, spectrometers and dosimeters) started in 2000. Since April 2003, Azerbaijan has participated in a programme to establish a computerized registry of radioactive material.

Ambient noise

The control of noise falls under the State Sanitary Epidemiological Service. At the *rayon* level, only the sanitary epidemiological centre in Baku is equipped with an acoustic laboratory accredited by the State Committee for Standardization to take noise measurements. This laboratory also measures electromagnetic fields in equipment and different environments. Noise control at airports and railways is the responsibility of the civil aviation and railway authorities, which have their own laboratories.

Noise measurements are taken in response to applications from companies and other entities that are seeking permission to start new activities, including building new houses. Measurements are also taken in response to complaints. The service provided by the acoustics laboratory is subject to the payment of fees by the applicants or plaintiffs. The laboratory may recommend measures to reduce noise (e.g. requesting traffic detours in the proximity of hospitals). The implementation of these recommendations is the responsibility of the competent sanitary epidemiological centre.

Traffic noise is measured only in response to complaints, which tend to be higher in summer, largely because windows are kept open in the summer months. According to experts from the acoustics laboratory in Baku, the number of complaints is decreasing because of improved insulation (doors and windows). Annoyance caused by noise from amusement activities is considered to be limited.

Noise standards for different environments and day times follow those issued by the Russian Federation (e.g. Sanitary Norm CH2.2.4/2.1.8.562–96 – Moscow, 1997).

Occupational health

Similar to what has been observed in other EECCA countries, Azerbaijan reported a dramatic decrease in deaths from occupational diseases to 0.33 deaths/100,000 population in 2001, a value that is about 5 times lower than averages reported in the EU. New cases of occupational diseases were in the range of 3.3 per 100,000 population, while 1.4 persons per 100,000 were injured in work-related accidents (this was 20 per 100,000 in 1990) In the past, occupational diseases arose mostly at the chemical facilities, few of which are now operating.

Although during the EPR mission it was not possible to acquire additional specific information, similarities with socio-economic changes that occurred in neighbouring countries make it highly plausible that the same explanations are applicable to Azerbaijan. On this assumption, the observed patterns are not the result of improved safety in the workplace, but rather a reflection of the profound changes in Azerbaijani society, where unofficial unemployment went up, in parallel with a rise in self-employment and informal economic activities, which escape official records. In addition, the meagre compensation available to disabled workers (e.g. in the form of pensions and compensatory leave) does not provide an incentive for workers to report occupational diseases. Finally, occupational diseases developed by workers in sectors where parallel health systems operate (e.g. army, civil aviation and railways) may not be included in the statistics of the Ministry of Health.

Landmines and unexploded ordnance (UXO)

Landmines and unexploded ordnance (UXO) are a problem of high environmental health significance, as they represent a major obstacle for the return of IDPs to their villages, and do not allow for safe use of the land for agriculture, pasture and other economic activities.

In the past 12 years, 400 people have been killed and 1,000 more injured by landmines and UXO. Although the 1994-1996 period was the most severe, new victims are still reported. In the past year alone, 11 people have been killed and 18 injured.

In an effort to deal with this internationally acknowledged "ecological threat", the Azerbaijan National Agency for Mine Action was established in 1998. It is responsible for setting priorities, planning, coordinating and managing all activities related to landmines and UXO. There are 100 million m^2 of mine-contaminated land, a legacy of the Nagorny-Karabakh conflict, and 6 million m^2 of land contaminated with UXO in the area close to the Baku-Tbilisi-Ceyhan pipeline construction site. A landmine impact survey was carried out

in early 2003. Its results will provide indications for future activities.

Landmines are found on the former battlefields of the Nagorny-Karabakh conflict, which has been an impediment to post-conflict reconstruction and rehabilitation. As the Government of Azerbaijan attempts to repatriate the IDPs, the land has to be inspected for human safety prior to reconstruction and rehabilitation. The mines were laid by unofficial military forces in an irregular manner, and there are no proper records of these mines. It costs US$ $1.50/m^2$ to search for and clear the mines. The United States' State Department is providing US$ 1.1-1.2 million per year. The goal is to clear the currently accessible area in 10 years and to clear the remaining 350 million m^2 (according to optimistic estimates) in 30-35 years' time.

The environmental problems resulting from UXO is mainly a legacy of former Soviet military operations. The Soviet army withdrew in 1991, leaving a number of military installations, which were only partially or inadequately dismantled. In May and June 2003, three people died from the explosion of UXO during dismantling. The expensive metal pieces inside the shells of UXO are trade items. The victims are poor rather than ignorant. An explosive ordnance disposal team has been established to dispose of UXO by explosion and dismantling with appropriate expertise. To this end €1.6 million have been allocated. The team consists of 40 experts and trained helpers.

12.3 Policy objectives and management

Policy framework

Azerbaijan inherited the health care system of the former Soviet Union, but its efficiency and effectiveness have deteriorated considerably, particularly at the primary level. Issues such as quality and access to medical services have become a subject of concern, conditioned, on the one hand, by the limited financing available to support the sector, and the deepening poverty of the population, on the other.

A WHO estimate suggests that public expenditure on health as a percentage of GDP declined from 2.1% in 1995 to 0.9% in 2000. The share of public expenditure in the total health expenditure declined from 77.7% in 1995 to 44.2% in 2000, while per capita total health expenditure in terms of purchasing power parity increased from US$ 49 to 57 in the course of six years.

The ongoing reform of the health sector has resulted in a partial shift to user fees in applicable sectors, and the development of private health care services. However, current health expenditures remain below pre-1990 budget allocations.

The drop in government expenditures has affected not only the health services, but also the availability and quality of other basic social services and infrastructure. Participatory studies show that rural areas and small towns suffer from unreliable supplies of energy and gas, declining infrastructure, as well as less access to basic health and education services.

Health sector reform was launched in 1998 with the establishment of a designated State Commission; and in 1999 a concept paper was adopted. The main thrust of health sector reform is currently on primary health care services, by improving outpatient treatment and prevention vs. inpatient treatment. The second main thrust is that of developing cost-effective health care services.

In line with the Millennium Development Goals, the strategic objectives include reducing the mortality rate of infants and children under 5, and in the maternal mortality rate, and monitoring access to safe drinking water. While the establishment of a regular and reliable water quality monitoring system is a key element, the reform does not extend to the sanitary epidemiological system, casting doubt on the effectiveness of a piecemeal approach disconnected from a broader reform of environmental health services.

To some extent, the impact of the environment on human health has been addressed as components of environmental initiatives such as the National Environmental Action Plan (NEAP). However, inter-agency coordination has been weak. This has undermined any sense of ownership and the commitment among the parties.

National environment and health action plan (NEHAP)

The draft NEHAP has been approved by the Ministry of Health and is now ready for consideration and likely adoption by the Government. It covers all aspects of environmental health, and some action items that it addresses are:

- The revision of national standards, including partial harmonization with international norms;
- Public information dissemination and awareness-raising;
- The introduction of new technologies and systems;
- The installation of new water and sanitation infrastructure.

The document also notes the significance of international waters for the country. The regional activities and action plan deriving from the Caspian Environment Programme, which includes five littoral countries, have been given prominence. The problems related to the contamination of the transboundary rivers Kura and Araz, which flow through Armenia and Georgia before reaching Azerbaijan, have also been highlighted.

In addition to the action topics that have been identified, the NEHAP states that financial support from international organizations and bilateral donors is necessary to carry out those actions needed to ensure a healthy environment and a healthy population.

State Programme on Poverty Reduction and Economic Development for 2003-2005

The State Programme on Poverty Reduction and Economic Development for 2003-2005 was adopted by presidential decree on 20 February 2003 to "reinforce social protection of the population for the near future and ensure a reduction in poverty in the country and implementation of the necessary measures to comply with the commitments of the 2000 United Nations Millennium Summit." The Ministry of Economic Development has been appointed to lead and coordinate its design and implementation.

The State Programme aims to work towards the Millennium Development Goals. It also refers to the Government's plan to increase expenditure on the health sector, which includes campaigns to reduce the incidence of TB, malaria, iodine deficiency, design public awareness campaigns for healthy lifestyles and nutrition, especially targeted to the poor, and monitor access to clean drinking water as well as other environmental factors affecting health. It also addresses the protection of public health by improving food safety through better control at State laboratories, improved coverage of veterinary vaccines, and upgrading quarantine control at the border. Improvement of health information systems, special consideration for refugees and IDPs, public investment in utilities and infrastructure measures, and rural development measures have also been included. The overall leadership for the implementation of the State Programme rests with the Ministry of Economic Development, and the Ministry of Health is directly responsible for only health-sector-related measures, which limits the participation of the public health sector in other relevant issues such as investment in water and sewerage infrastructure.

International agreements

Azerbaijan ratified the *Protocol on Water and Health* to the 1992 Convention on the Protection and Use of Transboundary Watercourses and International Lakes in January 2003. By becoming a Party to this international legal instrument for the fight against water-related diseases, the Government is expected to place emphasis on tackling the human health problems related to water in an integrated way and in an international context that involves the protection of water quantity and quality at all stages of extraction, treatment, distribution and disposal. However, the Government's financial commitment to this endeavour remains unclear.

Together with other States participating in the Third Ministerial Conference on Environment and Health (London, 1999), Azerbaijan adopted the *Charter on Transport, Environment and Health,* which sets targets, principles and a plan of action to move towards transport sustainable for health and the environment.

The Charter's objectives are being implemented through the Transport, Health and Environment Pan-European Programme jointly with UNECE. However, reliance on external support to ensure continued participation in the process and a lack of institutional mechanisms promoting cross-sectoral cooperation among relevant ministries represent a challenge to the Programme's effective implementation.

Legal framework

The Constitution provides for the "right to live in a healthy environment" (art. 39), the "right to health protection" (art. 41) and "environmental protection" (art. 78). The Laws on Environmental Protection and Environmental Safety, which were adopted on 8 June 1999 and together form the backbone of environmental legislation, mainstream human health in line with article 39 of the Constitution.

The key laws on environment and health in Azerbaijan are the Law on Sanitary-Epidemiological Services adopted on 10 November 1992, which defines the function and operation of the Ministry of Health and the State and regional sanitary and epidemiological services, and the Law on Public Health adopted on 26 June 1997.

Other laws, codes and decrees that relate to environment and health include the Law on Air Protection (27 March 2001), the Water Code (26 December 1997), the Law on Radiation Safety of the Population (30 December 1997), the Law on Food Products (18 November 1999) and the Resolution of the Cabinet of Ministers on Obligatory Medical Diagnostics for Workers (5 May 1998).

The authorization procedure for new projects and the issuing of "construction" or "technological passports" are similar to an environmental health impact assessment. It requires the involvement of specialists from the State Service to assess the compliance of new projects with sanitary norms. For relatively small projects, this is done by specialists from the *rayon's* sanitary epidemiological centre. It is, however, unclear according to which criteria the distinction is made.

The health assessment is carried out independently from the Ministry of Ecology and Natural Resources, which also plays a role in granting permits for new projects. The risk of overlap with assessments by the Ministry of Ecology and Natural Resources is judged to be low, as the two Ministries have different responsibilities and address different aspects of the potential issues posed by a new project.

The authorization of new projects proposed by international companies is subject to negotiation with the Cabinet of Ministers (chaired by the Prime Minister), which establishes the kind of documents needed for the project to be approved. This mechanism should allow for decisions to be taken on the basis of integrated information.

Institutional framework

<u>State Sanitary Epidemiological Services</u>

Azerbaijan inherited the traditional Soviet health service model, within which the Sanitary Epidemiological Service operates, governed by the Law on Sanitary Epidemiological Services. The Sanitary Epidemiological Service has two basic functions: (a) control and regulation of food, drinking and recreational water; noise, radiation and a variety of products; and (b) control of infectious and parasitic diseases.

Under the State Sanitary Epidemiological Service of the Ministry of Health come the State centres, which have specific functions and specialized laboratories. At the *rayon* level there are regional centres with specialized facilities including bacteriology, parasitology, virology and environmental laboratories, though the quality and capabilities of these facilities is not the same in all centres.

Defined by the Law, the main responsibility of the Sanitary Epidemiological Inspectorate is to protect the country from infectious diseases and other hazards posed by imported products. The Inspectorate is staffed by medical doctors and operates through eight stations at customs posts, where imported products are inspected.

Imported food, cosmetics, chemicals and detergents are inspected, as specified by the Law. Pesticides and pharmaceuticals, however, are the responsibility of the Ministry of Agriculture and a dedicated department of the Ministry of Health, respectively.

In addition, both local sanitary epidemiological centres and inspectors of the Ministry of Ecology and Natural Resources have the authority to inspect, for instance, factories, water-treatment plants and sewage systems, with the risk of duplication or, worse, inconsistent findings.

The State Sanitary Epidemiological Service is the main body with responsibility for monitoring the chemical and microbiological quality of drinking and recreational water, food and non-food products of domestic or imported origin (e.g. cosmetics, detergents, chemicals, material used for the production of pipes), soil, and radioactivity. It is also responsible for issuing certificates of compliance with relevant standards for new products. Its

specialists also train personnel working in the regional centres and perform quality checks on analyses carried out there. In addition, rapid investigation teams of epidemiologists may be dispatched to *rayons* to assist local epidemiologists in the investigation and containment of outbreaks of diseases and epidemics.

The Ministry of Health no longer monitors air pollution from transport and ambient noise, but this has been compensated by the emerging role that the Ministry of Ecology and Natural Resources has assumed with respect to air quality monitoring. Its results are accessible to the Ministry of Health, although the air monitoring system is not primarily designed to assess the health effects. Similarly, the relation between air pollution and certain health effects is not being investigated.

Based on agreements in the Commonwealth of Independent States, the monitoring of food, water, noise and other products complies with the technical norms issued by the Russian Federation and adopted by the Azerbaijan Government upon approval by the Scientific Research Institute, although they are adjusted to take into account local specificities.

At the *rayon* level, most laboratories are equipped to perform basic chemical-bacteriological analysis of sampled products. They can refer to the State laboratories to confirm analyses.

The Scientific Research Institute was established about 80 years ago and has an impressive track record, including a Nobel Prize awarded to one of its former Directors. The Institute has a staff of 300, of whom 108 are researchers. The Institute provides training and is responsible for the development of technical norms and standards for epidemiology and monitoring of food, drinking water and recreational water quality. The Ministry of Health is at the moment its only provider of financial support for research. Publication of scientific reports by the Institute is subject to prior approval by the Scientific Coordination Department of the Institute, as well as by the Scientific Committee of the Ministry of Health and by the Deputy Prime Minister.

Environmental health information systems

The main objective of the health information system is to provide the Government with the evidence to develop health-related policies. The sanitary epidemiological centres at the *rayon* level are the primary source of health-related statistics, which are compiled and analysed centrally by the Bureau of Health Statistics at the Ministry of Health. Information on notifiable infectious diseases is collected monthly, on immunization quarterly, and on the quality of food, water and other products annually. Outbreaks and epidemics are reported as they occur, to allow for prompt investigation and intervention.

The quality and completeness of the data collected at the *rayon* level may be partly affected by the difficulty of health personnel to adapt to disease coding according to the International Classification of Diseases (ICD), partly by the accumulation of errors in the recording and transcription of data, which are still done largely manually, and partly (as far as morbidity statistics are concerned) to under-reporting, because of informal practices and/or referral to private health care providers.

Food, water and other products subject to controls are monitored by the laboratories in the cities and regions in accordance with the country's standards and norms, and usually monthly or bimonthly. In the event of an outbreak or a situation posing an unusual risk (e.g. contamination of drinking water), daily monitoring is carried out. Similarly, monitoring of coastal recreational waters is intensified during summer. Azerbaijan does not currently have an information system integrating environment with health information.

In addition to the health system of the Ministry of Health, there are other independent systems run by different administrations (see chapter 3, on environmental information and public participation). The statistics reported by the Ministry provide an incomplete picture of the health situation in Azerbaijan. The Statistical Yearbook for Health of 2003 will for the first time also include data from these parallel health systems.

Since February 2003, there have been two major changes in the mortality and health-related data flows and classification, which may affect national health statistics in the years to come:

- Birth and death certificates will be routed directly to the Bureau for Health Information and Statistics of the Ministry of Health.

- The classification of the causes of mortality and morbidity has been changed from the 9th to the 10th Revision of the International Classification of Diseases (ICD10) and the National Centre for Health Monitoring has been established. The Centre will run registries for births and deaths, TB, HIV/AIDS and cancer.

The above changes will certainly represent a challenge for the country's health information system. A survey to identify data quality issues related to the new coding and reporting systems is planned. In addition, in the context of a health reform project, five *rayons* will serve as pilot sites for the implementation of the ICD10 classification.

In addition to health statistics available through the WHO Health for All Database, some selected and aggregated figures on health and demographic statistics can be accessed through the web site of the Ministry of Health (http://www.mednet.az/), and through that of the Trans-Caucasus Health Information Project.

Although these web sites provide information on some of the key health indicators, they do not allow for any analysis or comparisons to be made, nor for correlations with data on environmental quality parameters to be established. More comprehensive sources of mortality and morbidity data (including information disaggregated at city or *rayon* level) are the reports published by the State Statistical Committee on health care and on demographic indicators and the Statistical Yearbook. However, these publications do not appear to be easily accessible, and few health care providers know about them.

12.4 Conclusions and Recommendations

Azerbaijan has a population of over 8 million and is expected to reach 9.5 million in 2015. Urbanization is expected to continue. While people's overall health status appears to have improved since the mid-1990s, some diseases such as tuberculosis, malaria and acute intestinal infections continue to pose a threat to the population. The Ministry of Health will continue to need support to procure equipment and chemical supplies for the disinfection and extermination of mosquitoes, to equip laboratories for diagnosis, to expand eucalyptus planting, to train medical staff and to monitor the post-intervention situation. Furthermore, recent events have alerted the public health sector to the fact that it is not prepared to deal with natural disasters. Azerbaijan still has high rates of poverty and a large population of refugees and IDPs. Environmental degradation, particularly in the former industrial cities, is a suspected contributor to death, illness and disability, the causes and extent of which are not properly recorded or monitored.

While the challenges persist, government expenditure has declined, affecting the availability and quality of basic social services and infrastructure. As a result, there are, for example, unsanitary housing, a low level of hygiene, and poor quality of water supply, which create favourable conditions for the spread of food- and water-borne intestinal infections. The insufficient quantity of clean water is of particular concern.

It appears that the country's critical environment and health concerns are related to air quality, water and sanitation, food safety and waste management. But, above all, the institutional capacity of the public health sector needs to be further strengthened urgently, particularly in inter-agency coordination, standard-setting, and data collection and information management.

As has been reflected in the current health reform, the present structures for environmental health are geared more towards control and response than prevention. In addition, the control and monitoring system appears to suffer from a lack of adequate infrastructure and laboratory facilities, thus making it impossible to assess the real magnitude of environmental degradation and disease burden. Yet such information is essential for effective policy intervention. Although important steps have been taken to improve several aspects of health-related data collection and management, there is substantial room for improvement in the quality, completeness and accessibility of health and mortality statistics.

Over the past few years some studies have been conducted as fact-finding for further actions – on nutrition, water supply and sanitation, to name a few. To avoid duplication of information gathering and to ensure that different entities have the same knowledge of issues, such information could be shared among a wider audience through better inter-agency coordination, which might even lead to collective actions. Moreover, inadequate coordination and the lack of common understanding of initiatives among stakeholders often lead to national programmes

with limited involvement of relevant parties both in design and in implementation, which causes little sense of ownership and duplicates efforts.

The forthcoming adoption of the NEHAP represents an important opportunity to develop cross-sectoral cooperation, identify synergies with other major initiatives, such as the State Programme on Poverty Reduction and Economic Development for 2003-2005 and NEAP, and direct resources and actions towards priorities relevant to both the health and the environment sectors.

Recommendation 12.1:
(a) The Ministry of Health should revise and update the NEHAP, which was drafted in 1991, to reflect the current situation.
(b) The Ministry of Health should then ensure that the NEHAP is adopted and implemented in collaboration with the Ministry of Ecology and Natural Resources, and other relevant agencies and stakeholders.
(c) In particular, the respective ministries should consider joint application for funds for priority actions under the NEHAP, NEAP and the State Programme on Poverty Reduction and Economic Development.

There is currently a risk of duplication and inconsistency by the inspection functions of two different bodies – the Ministry of Health's Inspectorate and the Inspectorate of the Ministry of Ecology and Natural Resources. It is important that clear lines of responsibility are drawn and specific terms of reference given to each Inspectorate.

Recommendation 12.2:
The Cabinet of Ministers should redefine and clarify the respective functions and responsibilities of the Ministry of Health's Inspectorate and that of the Ministry of Ecology and Natural Resources. Such clarification should include their specific areas of competence, the conditions for intervention and their relation with local agencies, such as local sanitary epidemiological centres, and the procedures for handling potential disagreements, should they arise.

On top of ensuring compliance with norms and technical standards, experts should also engage in a critical appraisal of the overall monitoring system, and its functions, with a view to ensuring that the system serves the protection of public health in the most cost-effective and efficient way, taking into account also the latest international developments in the area. Improvement in this field would entail a thorough review of existing norms and standards and an in-depth revision of the monitoring procedures and the possibility to establish links between data related to the quality of water, food and other environment-related conditions with data coming from epidemiological and morbidity surveillance systems.

In addition, the rationalization of the monitoring system (e.g. by reducing the number of parameters to be monitored to those which are most relevant for public health protection) would help to concentrate resources on building capacity, upgrading monitoring and laboratory equipment, and improving analytical capacity only in selected areas.

Recommendation 12.3:
(a) The Ministry of Health, in cooperation with the Ministry of Ecology and Natural Resources, should develop a strategy for the overall monitoring of environmental samples and disease surveillance that enables an evidence-based approach to associating environmental status with impact on human health. This should be carried out in collaboration with WHO and other international organizations or bilateral donors to ensure coherence with international standards and practices.
(b) In particular, national legislation on quality assurance standards should be reviewed and adjusted, and existing overlaps and duplications, e.g. in relation to environmental monitoring responsibilities, should be assessed and removed (e.g. in air quality monitoring).

There is a need to improve the monitoring laboratories' analytical capacity and their adherence to quality assurance standards. At the same time, there are needs to upgrade the equipment of State and local laboratories, and to secure funds for the maintenance of this equipment. Consideration should be given to the possibility of partially recovering the cost of laboratory upgrades and equipment maintenance by providing value-added services, such as high-quality and sophisticated analyses, for a fee (e.g. to companies that may find it attractive to have their products analysed or certified without investing capital in developing in-house facilities to that end). This could also lead to the establishment of some analytical centres of excellence, where the initial investments in analytical equipment would ensure that the facility is used at full capacity and that the

staff become highly competent. National monitoring and surveillance systems would be strengthened, and laboratory confirmation levels of clinical samples would be improved. This can be achieved by training the experts of the sanitary epidemiological services.

Support for the work of the National Department of Environmental Monitoring would go a long way to improving public health provided that internationally standardized monitoring equipment is in place with a view to linking its findings with both ecological and health data, and orienting the service towards surveillance of priority environment-related diseases. For example, while reductions in the rates of acute intestinal infections remain a challenge, a major effort should go into improving their detection by strengthening laboratory capacity.

Recommendation 12.4:
(a) The Ministry of Health should revise the health information system in the light of the policy objectives to be achieved and of the supportive analysis to be performed.
(b) The Ministry of Health should develop indicators and establish and maintain rigorous procedures to ensure quality control and inter-laboratory comparability of results. The Sanitary Epidemiological Service could play a central role in developing and making available this capacity to local laboratories. It should also assess the possibility of developing partnerships with donors (e.g. international development agencies, foundations) to finance better laboratory facilities and technical capacity.
(c) The Ministry of Health should continue to direct major efforts towards building the appropriate infrastructure and capacity in health professions dealing with the primary collection and management of health statistics. This should be carried out in line with the above recommendation, and to the extent possible within the framework of international collaboration and support. High priority should be given to investing in a transition from a manual to an electronic system for the collection, storage, transmission and processing of health data.
(d) The possibility of developing partnerships and agreements with other key bodies, such as the Ministry of Ecology and Natural Resources, should also be considered for sharing information.

The experience of WHO with the development of a core set of indicators for environmental health monitoring could represent a useful starting point to map out data requirements and their sources, and to assess feasibility issues related to the implementation of such a system.

Research in environment and health-related matters would greatly benefit from greater interactions with the international scientific community, including for the development of possible partnerships and the identification of resources for strengthening the capacity of researchers in preparing robust research proposals, addressing relevant research questions, conducting and managing research, and presenting its results according to established international quality standards and procedures.

Recommendation 12.5:
The Ministry of Health should encourage and support the Scientific Research Institute in strengthening its international outreach and capacity to build partnerships for conducting and funding research. The submission of research results to scientific peer-reviewed international journals should be strongly encouraged, as should the identification of potential international partners and donors to support research activities. This should be accompanied by further developing researchers' professional skills, including through the development of exchange programmes with other scientific institutions.

The issue of safe water supply and adequate sanitation remains a challenge, both in Baku and in the rest of the country, and poses a major threat to health through increased risks of water-related diseases. The fact that only the most affluent can afford bottled water or filters also raises issues of social equity. The Ministry of Health could play an important role in advocacy and development of preventive strategies, in addition to maintaining its responsibilities in the control of drinking and recreational water quality.

Recommendation 12.6:
The Ministry of Health should take advantage of opportunities provided by being a Party to the

Protocol on Water and Health to develop partnerships with other relevant ministries and bodies and advocate the implementation of the policy recommendations set out in the Protocol, with a view to developing a comprehensive approach to water supply and sanitation, i.e. source protection, treatment and distribution of water; and disposal of human waste and waste water.

Radioactive contamination by low specific activity scales in residential areas in the vicinity of oil fields is raising concerns of a possible increased risk of leukaemia and other ionizing radiation-related diseases in the local population, and in particular in children. These concerns, however, have not yet been fully investigated.

Recommendation 12.7:
The Ministry of Health, e.g. through the Scientific Research Institute, and with WHO assistance, should support the efforts of the Radiation Medicine Department in investigating the possible health effects resulting from exposure to radioactivity from low specific activity (LSA) scales in residential areas in the vicinity of oil field.

In spite of some localized initiatives, which are aimed at the separate collection and incineration of medical waste, in the vast majority of urban and rural areas medical waste is disposed of together with municipal waste, potentially causing microbiological and chemical contamination.

Recommendation 12.8:
The Ministry of Health should work with the Ministry of Ecology and Natural Resources to revise present practices for the safe disposal of medical waste. Positive experiences developed in some health facilities (e.g. the separate collection of sharp materials in some hospitals in Baku) should be extended. The use of safe incinerating units should also be considered, as an alternative to landfilling, and criteria for the selection and operation of safe incinerators should be developed based on experience gained from existing programmes.

ANNEXES

ANNEX I

SELECTED REGIONAL AND GLOBAL ENVIRONMENTAL AGREEMENTS

	Worldwide agreements	Azerbaijan	
Year		Year	Status
1951	International Plant Convention	14.03.2000	R
1971	(RAMSAR) Convention on Wetlands of International Importance especially as Waterfowl Habitat 1982 (PARIS) Amendment 1987 (REGINA) Amendments	18.07.2001	R
1972	(PARIS) Convention on the Protection of the World Cultural and Natural Heritage	06.12.1998	R
1972	(LONDON) Convention on the Prevention of Marine Pollution by Dumping of Wastes and Other Matter 1978 Amendments (incineration) 1980 Amendments (list of substances)	22.04.1997	R
1972	International Convention on the International Regulations for Preventing Collision at Sea	22.04.1997	R
1973	(WASHINGTON) Convention on International Trade in Endangered Species of Wild Fauna and Flora 1983 (GABORONE) Amendment	23.06.1998	R
1985	(VIENNA) Convention for the Protection of the Ozone Layer	31.05.1996	R
	1987 (MONTREAL) Protocol on Substances that Deplete the Ozone Layer	31.05.1996	R
	1990 (LONDON) Amendment to Protocol	31.05.1996	R
	1992 (COPENHAGEN) Amendment to Protocol	31.05.1996	R
	1997 (MONTREAL) Amendment to Protocol	18.07.2000	R
1989	(BASEL) Convention on the Control of Transboundary Movements of Hazardous Wastes and their Disposal 1995 Ban Amendment 1999 (BASEL) Protocol on Liability and Compensation	01.06.2001	R
1992	(RIO) Convention on Biological Diversity 2000 (CARTAGENA) Protocol on Biosafety	14.03.2000	R
1992	(NEW YORK) Framework Convention on Climate Change	10.01.1995	R
	1997 (KYOTO) Protocol	18.07.2000	R
1994	(PARIS) Convention to Combat Desertification	24.04.1998	R

S = signed; R = ratified; D = denounced.

Selected bilateral and multilateral agreements *(continued)*

	Regional and subregional agreements	Azerbaijan	
Year		Year	Status
1957	(GENEVA) European Agreement - International Carriage of Dangerous Goods by Road (ADR)	18.07.2000	R
	European Agreement Concerning the International Carriage of Dangerous Goods by Road (ADR) Annex A Provisions Concerning Dangerous Substances and Articles Annex B Provisions Concerning Transport Equipment and Transport Operations		
1979	(BERN) Convention on the Conservation of European Wildlife and Natural Habitats	28.10.1999	R
1979	(GENEVA) Convention on Long-range Transboundary Air Pollution	02.04.2002	R
	1984 (GENEVA) Protocol - Financing of Co-operative Programme (EMEP)		
	1985 (HELSINKI) Protocol - Reduction of Sulphur Emissions by 30%		
	1988 (SOFIA) Protocol - Control of Emissions of Nitrogen Oxides		
	1991 (GENEVA) Protocol - Volatile Organic Compounds		
	1994 (OSLO) Protocol - Further Reduction of Sulphur Emissions		
	1998 (AARHUS) Protocol on Heavy Metals		
	1998 (AARHUS) Protocol on Persistent Organic Pollutants		
	1999 (GOTHENBURG) Protocol to Abate Acidification, Eutrophication and Ground-level Ozone		
1991	(ESPOO) Convention on Environmental Impact Assessment in a Transboundary Context	01.02.1999	R
1992	(HELSINKI) Convention on the Protection and Use of Transboundary Waters and International Lakes	18.03.2000	R
	1999 (LONDON) Protocol on Water and Health	21.10.2002	R
1994	(LISBON) Energy Charter Treaty	06.06.1997	R
	1994 (LISBON) Protocol on Energy Efficiency and Related Aspects	06.06.1997	R
1998	(AARHUS) Convention on Access to Information, Public Participation in Decision-making and Access to Justice in Environmental Matters	09.11.1999	R

S = signed; **R** = ratified; **D** = denounced.

ANNEX II

SELECTED ECONOMIC AND ENVIRONMENTAL DATA

Azerbaijan: Selected economic data

	1995	2000
TOTAL AREA (1 000 km^2)	86.6	86.6
POPULATION		
Total population, (1 000 000 inh.)	7.7	8.1
% change (1995-2000)		5.2
Population density, (inh./km^2)	90	94
GROSS DOMESTIC PRODUCT		
GDP, (billion US$)	2.4	5.3
% change (1995-2000)		120
per capita, (US$ 1000/cap.)	0.5	0.6
INDUSTRY		
Value added in industry (% of GDP)	27.3	36.0
Industrial production - % change (1995-2000)		32.0
AGRICULTURE		
Value added in agriculture (% of GDP)	25.2	15.9
ENERGY SUPPLY		
Total supply, (M toe)	13.2	11.7
% change (1995-2000)		-11.0
Energy intensity, (Toe/US$ 1000)	4.5	2.9
% change (1995-2000)		-35.6
Structure of energy supply, (%)		
Solid fuels		
Oil	26.6	
Gas	25.5	
Nuclear		
Hydro,etc.		

Source: UNECE and National Statistics
.. = not available.   - = nil or negligible.

Azerbaijan: Selected environmental data

	1995	2000
LAND		
Total area (1 000 km^2)	86.6	86.6
Major protected areas (% of total area)		
Nitrogenous fertilizer use (t/km^2 arable land)		
FOREST		
Forest area (% of land area)	11.5	11.5
Use of forest resources (harvest/growth)	53.7	63.8
Tropical wood imports (US$/cap.)		
WATER		
Water withdrawal (million m^3/year)	13,970.0	11,112.0
Fish catches (% of world catches)		
Public waste water treatment		
(% of population served)	31.6	31.6
AIR		
Emissions of sulphur oxides (kg/cap.)	6.5	4.4
" (kg/US$ 1000 GDP)	17.3	8.6
Emissions of nitrogen oxides (kg/cap.)	0.4	
" (kg/US$ 1000 GDP)	1.1	
Emissions of carbon dioxide (t/cap.)	4.0	3.5
" (ton/US$ 1000 GDP)	2.1	1.4

Source: UNECE and National Statistics
.. = not available.   - = nil or negligible.

SOURCES

Personal authors

1.   Akberov, M. The general information on natural resources of the Azerbaijan Republic and on scales of environmental problems
2.   Amaral, C. Proposal for institutional reform of the institutions dealing with rural development and sustainable use of natural resources, A concept and a road map
3.   Budagov, B. et al. The current state and causes of desertification of the Azerbaijan coastal area of the Caspian Sea. Conference on the issues of drought and desertification in the countries of south Caucasus. Tbilisi 2002.
4.   Kosayev, E. and Guliev, Y. Country Pasture/Forage Resource Profiles, Azerbaijan, FAO Grassland and Pasture Crops. 2001.
5.   Krever, V. et al. Eds. WWF. Biodiversity of the Caucasus Ecoregion. Baku – Erevan – Gland – Moscow – Tbilisi, 2001.
6.   Libert, B. CAB International. The environmental heritage of Soviet agriculture. 1995.
7.   Schmidt and Tirado eds. WHO Surveillance Programme for Control of Foodborne Infections and Intoxications in Europe, Seventh report 1993-1998.

Material from Azerbaijan

8.   Azerenerji Joint Stock Company. Azerbaycan Enerji Sistemi. Power System of Azerbaijan. Baku, 2002.
9.   Republic of Azerbaijan. State Programme on Poverty reduction and Economic Development 2003-2005. Poverty Reduction Strategy Paper (PRSP) of Azerbaijan. Baku 2003.
10.   State Committee on Ecology and Control of Natural Resources Utilization. National Environmental Action Plan. Baku, 1998.
11.   State Statistical Committee of Azerbaijan Republic and United Nations Population Fund. Demographic Indicators of Azerbaijan. 2002.
12.   State Statistical Committee of Azerbaijan Republic. Azerbaijan in XX Century. Part I. 2001.
13.   State Statistical Committee of Azerbaijan Republic. Azerbaijan in XX Century. Part II. 2001.
14.   State Statistical Committee of Azerbaijan Republic. Environment. Statistical Yearbook. 2002.
15.   State Statistical Committee of Azerbaijan Republic. Health Care. Statistical Yearbook 2002.
16.   State Statistical Committee of Azerbaijan Republic. Statistical Yearbook of Azerbaijan 2002.
17.   State Statistical Committee. The 2001 Household Budget Survey.
18.   The Constitution of the Republic of Azerbaijan. November 1995.
19.   The Law of the Azerbaijan Republic on Environment Protection. 1999.
20.   The Law of the Azerbaijan Republic on Environmental Safety. June 8, 1999.
21.   The Law of the Azerbaijan Republic on Statistics. 18 February 1994.
22.   The Republic of Azerbaijan. National Implementation of Agenda 21. Country profile. Baku, 2002.

Regional and international institutions

23.   Asian Development Bank with Japanese Fund for Poverty Reduction. "Grant Assistance to Asian Countries in Transition for Improving Nutrition for Poor Mothers and Children"
24.   Caspian Environment Programme. Regional Action Plan for Protection of Caspian Habitats. RER/98/G32/A/1G/31.
25.   Chemonics International Inc. Biodiversity Assessment for the Republic of Azerbaijan.
26.   Council of Europe. National report concerning the implementation of Ecological Network of Azerbaijan Republic. 2001.
27.   Economist Intelligence Unit. Azerbaijan. Country Profile 2002.
28.   Economist Intelligence Unit. Azerbaijan. Country Report. September 2002.
29.   Food Security in Azerbaijan 2001.
30.   Tacis. Assessment of Pollution Control Measures Azerbaijan. May 2000.
31.   UNDP. Azerbaijan Human Development Report 2002. Baku, 2002.
32.   UNDP. Grida. State of the Environment Azerbaijan. 1996.
33.   UNDP. Human Development Report 2003.
34.   UNDP. Living Conditions in the Azerbaijan Republic.
35.   UNECE. Towards a Knowledge-based Economy. Azerbaijan – Country Readiness Assessment Report. Geneva and New York 2003.
36.   UNEP. Grid Arendal. Caspian environment 2002. (CD-rom).
37.   United Nations. Azerbaijan. Country Profile. Johannesburg Summit 2002.

38.   USAID. Bureau of Europe and NIS. Biodiversity Assessment. 1999.
39.   World Bank. Azerbaijan Water Supply and Sanitation: Sector Review and Strategy. Report No. 20711-AZ. June 30, 2000.
40.   World Bank. Preliminary Assessment of Wood Production and Marketing in Azerbaijan. 2002.
41.   World Bank. Social Assessment of the Azerbaijan National Environmental Action Plan. Paper Number 32. Washington, July 1999.
42.   World Health Organization. Regional Office for Europe. European Health for All Database. June 2003.
43.   World Health Organization. Regional Office for Europe. Highlights on Health in Azerbaijan. March 2001.

Internet addresses:

Ministries and government institutions

44.   Azerbaijan Constitution: http://www.president.az/azerbaijan/const.htm
45.   Azerbaijan international. Impact of the war on the Environment.:
        http://www.azer.com/aiweb/categories/magazine/23_folder/23_articles/23_warenvironment.html
46.   AzUEIP: http://www.ecology-piu.org/
47.   Ministry of Economic Development: Privatization: http://www.economy.gov.az/HTML/Privatization/index.htm

Other internet sites

48.   AETC Azerbaijan Environment Technology Centre: http://www.rsk.co.uk/comprof/overseas.htm
49.   Atlapedia.: http://www.atlapedia.com/online/countries/azerbaij.htm
50.   Azerbaijan Center for the Protection of the Birds: http://www.azeribirds.org/eng/e_ekol_monitor.html
51.   Caspian Biodiversity Network: http://www.caspianenvironment.org/biodiversity/bmenu2.htm
52.   Caspian Environment Program: http://www.caspianenvironment.org/report_technical.htm
53.   Caspian Environment Programme. Coastal Profiles: http://www.caspianenvironment.org/itcamp/azeri4_2.htm
54.   Caspian Environment Programme. Regional Action Plan for Protection of Caspian Habitats:
        http://www.grida.no/caspian/additional_info/habitat.pdf
55.   Caspian Environmental Program Download library: http://www.caspianenvironment.org/report_technical.htm
56.   CIA Factbook: http://www.cia.gov/cia/publications/factbook/geos/aj.html
57.   Cleaning up in Azerbaijan. Article: http://permanent.access.gpo.gov/lps3997/9603azer.htm
58.   Council of Europe: http://www.coe.int/
59.   Development Gateway Azerbaijan: http://www.developmentgateway.org/
60.   Ecocaucasus org.: http://ecocaucasus.org/en/res.htm
61.   EIA. Country Analysis Brief: http://www.eia.doe.gov/emeu/cabs/azerbjan.html
62.   Friends of the Earth: http://www.foei.org/biodiversity/baku.html
63.   Gopa: http://www.gopa.de/en/projects/nr/natrsrc/czm/caspian_coast.html
64.   Governments on the Web: http://www.gksoft.com/govt/en/az.html
65.   Grida. Caspian maps and graphics: http://www.grida.no/caspian/maps_graphics/maps_graphics.htm
66.   Grida. Environmental Information System: http://www.grida.no/enrin/htmls/azer/azerb.htm
67.   Grida. Pollution loads: http://www.grida.no/caspian/priority_issues/env_quality/load.pdf
68.   Grida: State of the Environment: http://www.grida.no/enrin/htmls/azer/soe/ecology/index.html
69.   GTZ and University of Bremen: http://www.lexinfosys.de/index.html
70.   Integrated Coastal management: http://icm.noaa.gov/country/azerbaijan.html
71.   ISAR: http://www.isar.org/isar/caspian/info.html
72.   Library of Congress. Country Study: http://lcweb2.loc.gov/frd/cs/aztoc.html
73.   Regional Ecological Portal: http://ecocaucasus.org/en/res.htm
74.   UN Agenda 21: http://www.un.org/esa/agenda21/natlinfo/
75.   UN cartographic section, New York: http://www.un.org/Depts/Cartographic/map/profile/azerbaij.pdf
76.   UN in Azerbaijan: http://www.un-az.org
77.   UN: Johannesburg Summit 2002. Azerbaijan Country Profile: http://www.un.org/esa/agenda21/natlinfo/wssd/azerbaijan.pdf
78.   UNDP: http://www.undp.org/gef/workshop/documents/azerbaijan-report.pdf
79.   UNDP Azerbaijan Human Development Report 2000: http://www.un-az.org/undp/HDR2000_en.pdf
80.   UNDP Azerbaijan Human Development Report 2002: http://www.un-az.org/undp/nhdr/index.html
81.   UNEP: http://alpha.unep.ch/intranet/Country files - map/Azerbaijan/Azerbaijan.htm
82.   UNEP: http://www.unep.org/bpsp/Agrobiodiversity/agrobiodiversity thematic/Russiacis.pdf
83.   UNESCAP. Environmental Statistics in Azerbaijan: http://www.unescap.org/stat/envstat/stwes-azerbaijan.pdf
84.   USAID: http://www.usaid.gov/regions/europe_eurasia/countries/az/azerbaij.htm
85.   World Bank Country Brief:
        http://lnweb18.worldbank.org/eca/eca.nsf/Countries/Azerbaijan/608415C5560A1F3585256C24006B95DE?OpenDocument

86.   World Bank National Environmental Action Plan:
        http://lnweb18.worldbank.org/ECA/ECSSD.nsf/ECADocByUnid/2B3EF4C435BDB1A785256ACB0055DD14?Opendocume
        nt&Start=1&Count=10
87.   World Bank. Azerbaijan: http://lnweb18.worldbank.org/eca/azerbaijan.nsf
88.   WWF: http://www.worldwildlife.org/wildworld/profiles/terrestrial/pa/pa1305_full.html

Conventions and Programmes

89.   Basel Convention. Hazardous waste: http://www.basel.int/
90.   Bern Convention: http://www.nature.coe.int/english/cadres/bern.htm
91.   CITES : http://www.cites.org/
92.   Convention on Biological Diversity: http://www.cites.org/
93.   Ramsar Convention on Wetlands: http://www.ramsar.org/
94.   United Nations Convention to Combat Desertification: http://www.unccd.int/main.php
95.   United Nations Framework Convention on Climate Change: http://unfccc.int/